What's happening in

Hatice Necla Keleş & Ahu Ergen (eds.)

What's happening in cyber space?

An interdisciplinary approach

PETER LANG

Bibliographic Information published by the Deutsche Nationalbibliothek The Deutsche Nationalbibliothek lists this publication in the Deutsche Nationalbibliografie; detailed bibliographic data is available online at http://dnb.d-nb.de.

Library of Congress Cataloging-in-Publication Data
A CIP catalog record for this book has been applied for at the Library of Congress.

BAU
Publications

The articles in this book are first examined by independent reviewers. Then, editors do the corrections, and the text is prepared for publication. The ideas and opinions stated in the articles do not reflect the ideas and opinions of the editors. All the opinions in the book belong to the authors, and they can neither be accepted as Bahcesehir University's nor the editors' official opinions.

Printed by CPI books GmbH, Leck
978-3-631-80867-2 (Print) 978-3-631-81645-5 (E-PDF)
978-3-631-81646-2 (EPUB) 978-3-631-81647-9 (MOBI)
DOI 10.3726/b16722

© Peter Lang GmbH
Internationaler Verlag der Wissenschaften
Berlin 2020
All rights reserved.

Peter Lang – Berlin · Bern · Bruxelles · New York · Oxford · Warszawa · Wien

This publication has been peer reviewed.

www.peterlang.com

Contents

Introduction 7

Hatice Necla Keleş
1. New Skills and Talents for the Cyber World 9

Mehmet Sıtkı Saygılı
2. Supply Chain and Logistics Management in Cyberspace 21

Ahu Ergen
3. Smart Retailing in Cyberspace 49

Erol Eryaşa
4. Cyberspace Impacts on Maritime Sector 69

Ahmet Naci Ünal
5. Cyberspace and Chaos: A Conceptual Approach to Cyber
   Terrorism 103

List of Figures 127

List of Tables 129

Introduction

The term "cyberspace" that entered into our lives towards the end of the 20th century is defined as: "a global platform that consists of the network containing the infrastructures of information technologies with Internet, communication networks, computer systems, embedded processors and controllers". When this term is analysed, it is seen that nearly all informatics systems minimize the human control and at many platforms, they are connected to each other. Although the term "cyberspace" is very new, it is like a living organism that is coordinated with scientific disciplines such as finance, defence, health, education, transportation and security, in all manned and unmanned areas.

Our interaction with this organism covers almost all our lifespan. When looked closer, it is seen that cyberspace must be researched not only from technology side but also from sociology and psychology sides.

The improvements in robotics and artificial intelligence increase the concern of high unemployment rates in the close future. However, many scientists state that these improvements may create new opportunities and employment types. Shortly, it is stated that today, where the jobs are converted into cooperations between man and machine, the only thing remains is to find ways to adapt cyberspace. These opinions are the subject of Chapter 1.

Technological improvements support business world and the digital transformation makes the business processes faster and more transparent. Especially, when supply chain and logistics management processes are considered, complex business processes get easier with cyber networks. This makes real-time data sharing, faster business processes, customer-focused working and low costs, possible. With all these facts, it is also important to provide security of the supply chains and logistics processes. All these related topics are covered in Chapter 2.

In dynamic competition, the use of cyberspace in retailing causes a vital transformation in the sector. Many modern retailers are using smart technologies in their retail stages to enhance consumer experience. Wearable technologies, mobile Internet, cloud and big data, IoT are among the most popular smart technologies in retailing. Chapter 3 covers these new trends both from retailer and consumer perspective with current technological innovations.

Maritime industry is keeping its important position in today's global economy. Especially in the digital transformation process of Industry 4.0, maritimes industry with infrastructures, ports, ships and companies that are managing the ships, is using all the opportunities of cyberspace. So, maritime industry is facing both high physical and cyber risks. Being unable to manage these risks and not increasing the cyber security awareness of human in this sector, may cause serious loss in the sector. What must be done to minimize these risks is the topic of Chapter 4.

Terrorism definition is a phenomenon for which the countries and international society cannot have a consensus. When terrorism's target is considered, it is agreed that a societal chaos is created to threat human life and lifestyle. Starting with the 21st century, the use of cyber technologies by terrorists for publicity, communication and terror activities caused to name the term "cyber terrorism". For the society to live in peace and security, the individuals who use cyberspace must be aware and the security forces must keep their strategies always up to date. Readers who would like to increase awareness with scientific approach may benefit from the Chapter 5.

So, welcome to the cyberspace journey that we prepared for you in our book *What's Happening in Cyberspace?*

Hatice Necla Keleş

1. New Skills and Talents for the Cyber World

Abstract: There are many uncertainties about the future of works in the cyber world and how developing technologies will be. Technological advances especially in robotics and artificial intelligence fields will lead to great unemployment. Such opinions are increasing gradually. However, technological advances can create new opportunities and forms of employment. It will be not only to destroy all the occupations of robots and technology but also to take over many tasks in the professions. In this case, the focus is on looking for ways to keep up with the reshaped cyber world where professions turn into partnerships between man and machine and to acquire new skills.

Keywords: Skill, Talent, Profession, Cyber World

1 Skills and Profession

After I graduated from the university, I was faced with the question of what I was going to do. I must admit that it is a late period to face such a question. I found myself in a foreign exchange department of a bank. The first day I started, everything started well. When lunchtime came, I asked the unforgettable question: *"Where is the staff canteen?"* Of course, there were none! The difference in the organizational structure of banks had a shocking effect for me. I admit that I was too late to find this out. As a young and newly graduated employee, it was necessary for me to be ready for the first working day. I spent unhappy three years at this bank. After three years, I realized that I could not continue working in such an environment that did not suit my abilities. After my decision, I have now left 20 years behind in my happy career in workplaces where there are canteen, garden, festivals and library full of students. It is not possible to deny the contribution of my three unhappy years in banking job. Firstly, it showed me how important it was to choose a profession. It also showed which jobs fit my abilities and skills. It took me three years to realize the career which fit my skills and abilities and learn how to proceed. I consider myself lucky when I see people who have to work in professions that do not fit their abilities all their lives. I could have realized it much later—like some people—and even I could have pursued a career without even realizing it. "I could have pursued a career" because it would not just continue on its own, I had to continue it.

Unfortunately, today, there are still employees in workplaces who do not realize the profession that fits their abilities. However, the thing that is more sorrowful is that there are children in schools who do not realize the field that fits their abilities. In the seminars I speak about the choice of professions that fit one's talents, I have the chance to meet parents who have children from different age groups. I say that their children are lucky to have such parents who have 15-year-old children and who are interested in talent and career choices. I also say that the 10-year-old children who have interested parents are luckier and 10-year-old children who have interested parents are much luckier. Because I believe the sooner the better.

2 "War for Talents"

If we consider what the profession is, it is a certain work which is acquired with a certain training, aims to produce useful goods and/or services for people, which is done to earn money in return, and which is determined by certain rules. The important point here is "to produce useful goods and/or services for people". In the historical process, while the action of "to produce useful goods and/or services for people" has not changed, "the ones to be produced" are constantly changing according to the needs of people. Today, when we think about this process, it is difficult to find a similarity of the needs of our great-grandparents with ours. So, how much will our grandchildren be similar to our needs after a decade or after 30 and even 50 years? For this reason, this issue is always on the agenda. There are people among us who have different jobs compared to the past, and the number of the jobs that did not exist in the past is increasing. The answer to the question "What will you become when you grow up?" is a huge question mark today. Because we are in a time where there are fundamental changes in how talent diversity happens.

With the transition from agricultural and industrial society into the information society we are currently living in, the "human" factor has started to be considered as an indispensable resource for businesses with a developing and changing perspective.

As a result of globalization and technological developments, the definition of the labor force started to differ. The human factor is considered as a talent that creates added value in the enterprise in the information age. Businesses that have talented employees and apply differentiation strategies successfully have continued to exist until our present day.

Since 1998, when a group of McKinsey advisors used the phrase "war for talents" and laid the foundations of it, the belief in the importance of talents gained importance for organizational excellence (Michaels, Handfield-Jones and Axelrod, 2001); and talent management became an increasingly popular field (Chuai, Preece and Iles, 2008). After McKinsey & Company used the term "war for talents" as the name of its original study on talent management applications and perceptions, it completed its second study in 2000, and updated the findings of the first one. In this short time period, a phenomenon, which was experienced by many people before but not fully named, was named, and since then, the expression "war for talents" has taken place in the business world (McKinsey & Company Inc., 2001: 1). It is seen that the talent shortage, which caused talent battles, is still continuing to increase with talent management practices (McKinsey & Company, Inc., 2001).

Contemporary English Dictionaries define "talent" as "natural aptitude or skill people possess" (Stevenson, 2010). The second meaning of talent found in contemporary English Dictionaries refers to a person or persons of talent (talent as subject), that is, people possessing special skills or abilities (Ashton and Morton, 2005). "Talent" has become one of the powerful words in the dictionary of businesses today. Each employee is considered as a talent, and keeping these talents in the business, ensuring their commitment and the development of new talents in businesses have begun to gain more and more importance (Peters, 2006: 12). If success is achieved through people in competition, it is becoming increasingly important to create a unique workforce for competitive advantage (Pfeffer, 1995:17). For the purpose of providing competitive advantage for businesses, the talent management concept has been created to develop themselves and reach their career goals by creating suitable settings for talented employees and developing a setting where talented employees will adapt themselves. The understanding of using employees as a resource has been replaced with the talent management concept. Businesses target to carry their organizations to the future by focusing on career and development planning of the talents that are acquired with talent management implementations.

Talents provide competitive advantage. We cannot explain this better than Nersesian, who is the president of Electronic Measurement Group of Agilent Technologies: "At the end of a day, all that remains is the improvement in the talents of people. All our products may disappear in time. All that remains is corporate learning and the improvement of skills and talents that exist in our people" (Conaty and Charan, 2011: 10).

In the 1960s, career management was considered as a way of helping individual employees in achieving their targets with applications like career planning workshops and counseling. Until 1980s, the business world had already changed, and career management had begun to be considered as a means that targeted to cover business needs with planning the replacement of resigning employees and the promotion programs for key employees. Towards 1990s, career management was considered as adopting the needs of the individual to the requirements of the business. Because age-related promotion opportunities were no longer realistic for businesses, and maximizing individual potential was considered as an important element in business success (Yarnall, 2008: 28). Today, the most critical factor in human resources is ensuring the employment, training, dynamization, motivation and long-term employment of talented employees. In today's business world, people face the ever-increasing global competition, changing markets and unexpected situations; and it is increasingly difficult to employ, keep, and improve talented employees that are required by businesses (McCauley and Wakefield, 2006: 4).

Technology is becoming dependent on "knowledge employees" who constitute the intellectual capital for developing new products and for applying effective marketing strategies (Santhoshkumar and Rajasekar, 2012: 39). Having talented employees and benefiting from them at the highest level are considered as the main source of being innovative, creating value, providing difference in competition and effective performance in companies (Gregoire, 2006: 6). Businesses need more talents and talent management because of its impact on economic power. Talents have important roles in having a more productive structure and becoming more recognized (Mucha, 2004: 96–100). According to Ashton and Morton

(2005), talent management is a strategic and holistic approach for human resources and business planning or for a new path for organizational effectiveness. Talent Management develops the performance and potential of people who can make a measurable difference within the organization. It also provides better performance within all levels of the labor force and enables everyone to reach their potentials. Talent Management has become one of the concepts that is associated with the importance that can be described as the driving force of businesses, and therefore, of markets.

In his book *The War for Talent*, Fishman (1998) described "Talent Management" as one of the effective factors of economic life. In the strategic model for talent management of Collings and Mellahi, it was aimed to evaluate the existing talents of talented staff and talent management to increase the business performance and to provide employee incentives and dependence on the business with the organizations that would be realized as a result of this evaluation (Collings, 2009: 304–313). Among the Talent Management processes, there are workforce planning, talent gap analysis, personnel selection, training and development, retention, talent evaluation, successive planning, and evaluation (McCauley and Wakefield, 2006).

It will not be possible to interpret the present and even foresee the future without knowing what happened yesterday. Many things have changed since yesterday. What has remained unchanged is "the necessity of meeting the human needs". In every period in the history of mankind, in every geographical area, in every culture, people have had requirements that needed to be covered. When a human being discovered that he could receive more than one seed if he planted it on the soil, the life changed completely after permanent settlement. In time, the changes in economy, politics and technology were accepted as industrial revolutions, and provided the basis for the beginning of many topics that are on the agenda today. The transition from feudal structure to the industrialization with the use of steam was the first step of successive developments. The changes in this respect meant factory, employee, division of labor, specialization and working environment. The basic role was again realized with "the necessity of meeting the human needs" in the shaping of the period. The revolution that took place in industrialization with "Fordism" paved the way for the Third Industrial Revolution with the introduction of programmable automation systems. New business lines, new expertise areas and new competencies and skills emerged with these developments. In the historical process, the name of "unmanned" production in the transformation of industrialization was made for the first time as "Industry 4.0" in a fair in 2011 in Germany.

We are more digital now. We are more robotic now. We are more electronic now. We consume everything quickly. It is predicted that future generations will have longer lifetimes. Both the decrease in infant mortality, and the doubling of human life expectation when compared to 1900s show that we will live more in the future. This means more consumption, more production, and longer careers. When it is considered that many professions will be defeated by automation at this point of industrialization, it seems that changing and transforming will be inevitable for each of us.

Ever-increasing technological, demographic and socio-economic changes is transforming industries and business models, changing the set of skills that are required by employers, and shortening the lives of the existing skills of employees (Chui et al., 2015). Even the jobs which are not affected by technology directly and that have more stable employment structure are changing, and may need very different set of skills after a few years. In this new environment, the business model change causes that the set of skills are also disrupted often almost simultaneously and with only a minimum time delay.

In previous industrial revolutions, it would take decades to build the education systems and labor market institutions that were required for developing new skill sets. However, this may not be a simple option when the approach speed and disruption scale brought by the Fourth Industrial Revolution are considered (Infosys, 2016). Much of the subject knowledge of current workforce will be outdated in just a few years. Employers are often concerned about the skills and competences that new employees will be able to use to successfully perform their tasks in addition to the difficult set of skills and formal qualifications. The decisions of employers using new technologies in recruiting new employees must be based not only on training but also on non-cognitive skills that enable employees to succeed in work-related learning (Bessen, 2014).

We have to learn new skills to survive in this world that is being reshaped again. This can be achieved by knowing the individuals and their abilities better. Knowing who you are, what you like, what you do best, how to

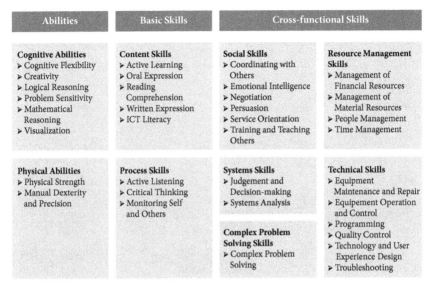

Abilities	Basic Skills	Cross-functional Skills	
Cognitive Abilities ➤ Cognitive Flexibility ➤ Creativity ➤ Logical Reasoning ➤ Problem Sensitivity ➤ Mathematical Reasoning ➤ Visualization	**Content Skills** ➤ Active Learning ➤ Oral Expression ➤ Reading Comprehension ➤ Written Expression ➤ ICT Literacy	**Social Skills** ➤ Coordinating with Others ➤ Emotional Intelligence ➤ Negotiation ➤ Persuasion ➤ Service Orientation ➤ Training and Teaching Others	**Resource Management Skills** ➤ Management of Financial Resources ➤ Management of Material Resources ➤ People Management ➤ Time Management
Physical Abilities ➤ Physical Strength ➤ Manual Dexterity and Precision	**Process Skills** ➤ Active Listening ➤ Critical Thinking ➤ Monitoring Self and Others	**Systems Skills** ➤ Judgement and Decision-making ➤ Systems Analysis **Complex Problem Solving Skills** ➤ Complex Problem Solving	**Technical Skills** ➤ Equipment Maintenance and Repair ➤ Equipement Operation and Control ➤ Programming ➤ Quality Control ➤ Technology and User Experience Design ➤ Troubleshooting

Source: World Economic Forum, based on O'NET Content Model.
Note: See Appendix A for further details.

Tab. 1.1: Core Work-Related Skills (http://www3.weforum.org; O'NET-ISCO Conversion Tables, http://www.bls.gov/soc/ISCO_SOC_Crosswalk.xls)

activate your brain, and knowing the things that will enable you to do your best are more important than ever in this business world that is being shaped again. I believe that there will be space for different competencies, skills, roles, and for everyone in the future. It appears that the ability to acquire future competencies, communication and coordination skills come to the forefront as strategic skills. Companies experience great difficulties in preparing their workforce for the new era in the time of faster changing skill requirements and new organizational structures. Companies do not foresee that machines are substitutes for people as kits. Instead, their focus is on building a workforce that has the right skills to complement new technologies and enable the company to use its powers. This will constitute a wider challenge for companies because it tries to create a "learning economy" where the skills of employees continue to develop and adapt to innovation in society (Stiglitz, 2012).

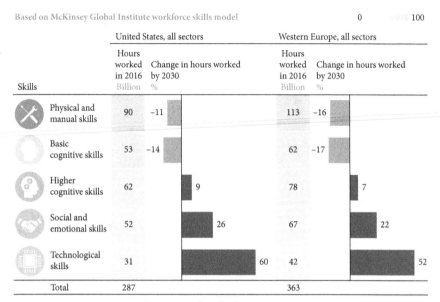

Skills	United States, all sectors			Western Europe, all sectors		
	Hours worked in 2016 Billion	Change in hours worked by 2030 %		Hours worked in 2016 Billion	Change in hours worked by 2030 %	
Physical and manual skills	90	−11		113	−16	
Basic cognitive skills	53	−14		62	−17	
Higher cognitive skills	62		9	78		7
Social and emotional skills	52		26	67		22
Technological skills	31		60	42		52
Total	287			363		

NOTE: Western Europe: Austria, Belgium, Denmark, Finland, France, Germany, Greece, Italy, Netherlands, Norway, Spain, Sweden, Switzerland, and the United Kingdom. Numbers may not sum due to rounding.

SOURCE: McKinsey Global Institute workforce skills model; McKinsey Global Institute analysis

Fig. 1.1: Workforce Skills Model (McKinsey, 2017).

Skills are a key problem for this period. A well-trained workforce that has the skills that are needed to adopt to automation and to adopt the AI technologies will enable that our economies will increase productivity and ensure that all employees use their capabilities. Not being able to deal with the demands of changing skills may exacerbate social tensions and cause increased skills and wage deviations—and create a divided community with people who are successfully used in the rewarding business and those with declining wage expectations in traditional jobs. Organizational and human resources effects are important for companies. It is not only a problem for companies. Policy makers, employee agencies, non-profit organizations, business associations and trade unions will work with business leaders to ensure that the conditions for skills development that will be needed are valid. The new condition of our automation period is the transition to the "learning economy" where human capital is the most important element. The future welfare of our societies and the future well-being of our workforce depend on whether or not we can achieve this goal (McKinsey Global Institute, 2017).

Studies show that businesses will change in four basic ways. Firstly, businesses will undergo a change of mentality: the key to their future success will be providing continuous learning options and imposing a lifelong learning culture throughout the lifespan of their businesses. Secondly, the basic organizational structure will change: there will be a stronger shift towards cross-functional and team-based work, for this, more agile ways of work with less hierarchy may be needed together with the creation of new business units. Thirdly, the appointment of business activities will be changed with "unrestricted" and "reassembling" means. This will ensure that companies (and especially large-scale practitioners) use their different competence levels of their workforce in the most effective way. Fourthly, the composition of labor will also change. Freelancers and other contractors will have more jobs that will increase the "concert" or "sharing" economy that emerges. To organize these changes, senior leadership and some functions will play key roles. CEOs and senior executives who will experience these challenges will need to adopt the right automation

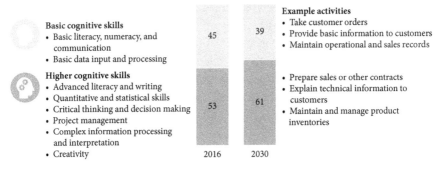

NOTE: Western Europe: Austria, Belgium, Denmark, Finland, France, Germany, Greece, Italy, Netherlands, Norway, Spain, Sweden, Switzerland, and the United Kingdom. Numbers may not sum due to rounding.

SOURCE: McKinsey Global Institute workforce skills model; McKinsey Global Institute analysis

Fig. 1.2: Workforce Skills Model Percentage of Time Spent on Cognitive Skills (McKinsey, 2017)

and AI mentality together with the information they need to direct change. Human resources departments will have to undergo profound changes in the way they work while skills and roles change and talents become more important (Willcocks, 2016; Jesuthasan and Boudreau, 2017). After the decision is taken for adopting automation and artificial intelligence technologies, leaders face difficult and obstinate questions on how to apply this decision (Jesuthasan, 2018):

1. How, when and where must automation be applied in our organizations? Is it a simple choice between people and machines?
2. As businesses and automation continue to develop, how do we stay at the top of these technological trends?

As for employees, those who can adapt to the changes brought by technology with education and training will find more satisfactory jobs in the future. The biggest difficulty business leaders will face is combining human, automated roles and responsibilities effectively, and offering human employees opportunities to practice their natural talents. Contrary to the worst fears of our present day, robots may facilitate the rise, not the death of such knowledge employees. However, managers will do well in ensuring that their employees do the necessary work, acquire expertise and train again when necessary to prepare for the inevitable changes in their existing job roles (Willcocks, 2016).

Ensuring that employees are trained again, and enabling them to learn new marketable skills in their lives are critical difficulties. Since the skill set that is required for a successful career had changed, re-training the middle career in accordance with talents will become even more important. Businesses may be leaders in some areas like in-service trainings and in providing opportunities for employees to develop their talents. Organizations must create a more agile working environment to support the transformation that is continuing and to think about how business levels will change. In a world where the roles and meanings of our jobs has begun to change, creative insights will be needed on how our lives will be organized and valued in the future; however, the starting point will continue to be the subject of talent in every period.

References

Ashton, C. and Morton, L. (2005). "Managing talent for competitive advantage". *Strategic HR Review*, 4(5), 28–31.

Bessen, James. (August 25, 2014). "Employers Aren't Just Whining – the 'Skills Gap' Is Real". *Harvard Business Review*, 14 August 25, https://hbr. org/2014/08/employers-arent-just-whining-the-skills-gap-is-real.

Chuai, X., Preece, D. and Iles, P. (2008). "Is Talent Management Just 'Old Wine in New Bottles'?: The case of multinational companies in Beijing". *Management Research News*, 31(12), 901–911.

Chui, M., Manyika, J. and Miremadi, M. (2015). "Four Fundamentals of Workplace Automation," McKinsey Quarterly. https://roubler.com/au/wp-content/uploads/sites/9/2016/11/Four-fundamentals-of-workplace-automation.pdf

Collings, D. G. and Mellahi, K. (2009). "Strategic Talent Management: A Review and Research Agenda". *Human Resource Management Review*, 19(4), 304–313.

Conaty, B. and Charan, R. (2011). *Yetenek sarrafları*. İstanbul: Kapital Medya Hizmetleri.

Fishman, C. (1998) *The War for Talent*. New York: Fast Company.

Gregoire, M. (2006). "Consistently Acquiring and Retaining Top Talent". *Workforce Management*, 85(19). http://www3.weforum.org/docs/WEF_Future_of_Jobs.pdf [24.09.2019]

Infosys. (2016). *Amplifying Human Potential: Education and Skills for the Fourth Industrial Revolution*. Bangalore, India: Infosys.

Jesuthasan, R. and Boudreau, J. (2017). "Thinking Through How Automation Will Affect Your Workforce." *Harvard Business Review*. https://hbr.org/2017/04/thinking-through-how-automation-will-affect-your-workforce [30.08.2019]

Jesuthasan, R. (August 8, 2018). "Future of Work—Reinventing Jobs". https://www.willistowerswatson.com/en-US/Insights/2018/08/future-of-work-reinventing-jobs

McCauley, C. and Wakefield, M. (2006). "Talent Management in the 21st Century: Help Your Company Find, Develop and Keep Its Strongest Workers". *The Journal For Quality & Participation*, Winter, 29(4), 4–7.

McKinsey & Company, Inc. (2001). *The War for Talent, Organization and Leadership Practice*. USA: Harvard Business School Press.

McKinsey Global Institute. (2017). San Francisco, California, www.mckinsey.com [26.08.2019]

Michaels, E., Handfield-Jones, H. and Axelrod, B. (2001). *The War for Talent*. Boston: Harvard Business School Press.

Mucha, R. T. (2004). "The Art and Science of Talent Management". *Organization Development Journal*, 22(4), 96–100.

Peters, T. (2006). "Leaders as Talent Fanatics". *Leadership Excellence*, November, 23(11), 12–13.

Pfeffer, J. (1995). *Rekabette üstünlüğün sırrı: İnsan*. 1. Baskı. İstanbul: Cem Ofset.

Santhoshkumar, R. and Rajasekar, N. (2012). "Talent Measure Sculpt for Effective Talent Management: A Practical Revise". *The IUP Journal of Management Research*, XI(1), 38–47.

Stevenson, A. (Ed.). (2010). *Oxford Dictionary of English* (3rd ed.). Oxford: Oxford University Press.

Stiglitz, J. (2012). "Creating a Learning Society". The Amartya Sen Lecture.

Willcocks, L. (2016). "Why Robots May Not Be Taking Your Job—at Least, Not in the Next 10 Years: How Organisations Can Embrace Automation". *European Business Review*, 15–17. www.europeanbusinessreview.com

Yarnall, J. (2008). *Strategic Career Management Developing Your Talent*. USA: Butterworth-Heinemann.

Mehmet Sıtkı Saygılı

2. Supply Chain and Logistics Management in Cyberspace

Abstract: Today, in business world, many valuable information is shared at any time during the operational processes in cyberspace. Developments in technology bring new opportunities for supply chain and logistics management. Especially, the digital transformation in business, powered by technological improvements, made the business life faster and more transparent. Using the cyber-physical systems, big data, Internet of things, smart factories have positive effect on the competitiveness of companies in supply chain. At the same time, real-time ERP systems, digital purchasing, smart warehouses and new operating processes in transportation get logistics services shortened and reduce costs. However, the security issue must also be taken into account due to destructive, disruptive and unpredictable effects.

Keywords: Supply Chain Management, Logistics, Cyberspace, Industry 4.0

The term "cyberspace" was first mentioned in William Gibson's *Burning Chrome* which was published in July 1982. The foundations of cyberspace were established by the US Advanced Research Projects Network (ARPANET) (Almagor, 2011: 48), which emerged in the 1960s to share information across computers in a network, and the Internet protocol was standardized in 1982 to send data packets over this network (Frenzel and Frenzel, 2004: 167). In 1989, the use of the Internet was created by the establishment of an information-sharing system called "World Wide Web (www)" in the laboratories of the European Organization for Nuclear Research (CERN) by Tim Berners-Lee, thus creating a virtual environment in which information systems and people constantly interact (Lee et al., 1992: 455). The results of the activities carried out in this environment can be physical and affect people with or without awareness. Such an effective virtual environment is called cyberspace and has taken its place in the history of humanity (Ünal and Ergen, 2018: 192).

Developments in information and communication technologies led to increasing connectivity opportunities in cyberspace. These opportunities play an important role in the development of intercompany cooperation and in the increase of total productivity in order to benefit from the global economy. The international supply chain and logistics flow realized by companies that can communicate on a global scale provides real-time connection with operation processes, shorter business processes, flexibility, decreased costs of stocking, and brings the products and/or services to market in a shorter time. In line with these benefits, all stakeholders in the supply chain are connected to each other by opening up information and communication technology infrastructures in cyberspace in order to act in real time, orderly, fast and transparently.

2.1 Digital Transformation in Supply Chain and Logistics Management

The traditional structure of the supply chain involves the supply of raw materials, semi-finished products and ready-to-use parts, followed by their conversion into products in a production environment, and then their distribution to the final consumer or industrial buyer. In this context, it is the whole of the relations and connections that provide material, product and information movement between suppliers, producers, distribution channels and consumers (Felea and Albastroiu, 2013: 75). Any misdesigned connection reduces the overall efficiency of the entire supply chain (Janvier-James, 2012: 195). The material, product and information flow in each stage of this structure are supported by logistics services including transport, storage, customs and insurance. It should be noted that global supply chains and logistics networks are the backbones of the global economy by supporting trade, consumption and economic growth (World Economic Forum, 2012: 4). Therefore, they should be correctly planned, organized, executed, coordinated and supervised.

The global digital transformation changes production processes, service models and interactions with other companies. Consumer demands are directed towards personalized products. Manufacturers use technologies to produce these products. Customers and business partners are integrated into business processes and logistics service providers offer solutions with new business models. As a result of these developments, it is difficult to choose useful and usable information out of the increasing information accumulation in cyberspace, thus increasing uncertainties. In order to reduce uncertainty in the management of supply chain and logistics processes, and to find the information needed, there is a need for auxiliary systems (Yarman and Ünal, 2015: 31). The Internet of Things (IoT), Cyber-Physical Systems (CPS), Big Data, Smart Factories, cloud computing, autonomous systems, augmented reality, three-dimensional (3D) and four-dimensional (4D) printing, especially with the fourth industrial revolution (Industry 4.0), are some of the technologies that are utilized as auxiliary systems. Companies focus on investing in these technologies in order to provide real-time information, speed, and flexibility and efficiency in resource planning, new product generation, purchasing, storage, transportation and distribution processes, while also keeping pace with digital transformation.

2.2 Industry 4.0

Industrial revolutions affect the development of international trade and economic relations. Until today, there have been four revolutions in industry. As seen in Fig. 2.1, there have been different developments in every industrial revolution. The first industrial revolution (Industry 1.0) was the mechanization of production processes using water and steam power in the 18th century. The second industrial revolution (Industry 2.0) was the introduction of mass production by utilizing electrical energy in the 19th century. The third industrial revolution (Industry 3.0) was the introduction of computer-controlled production systems and production lines of programmable machines in the 20th century. Today, under the fourth industrial revolution called Industry 4.0, intelligent machines and

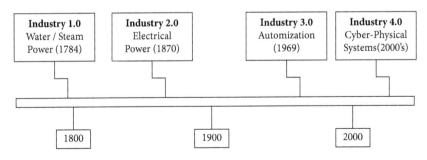

Fig. 2.1: Stages of the Industrial Revolutions (Boulila, 2019: 20)

artificial intelligence are used to integrate virtual and physical systems (Ünal and Saygılı, 2019: 3).

The concept and scope of Industry 4.0 was first presented at the 2011 Hannover Fair in Germany (Öztürk, 2017: 17373). Germany is among the ten projects under the "Future Projects" program under the "High Technology Strategy-2020 Action Plan" for achieving innovation targets over a period of ten years (European Commission, 2017: 3). With the new industrial revolution, there is a transformation towards production systems where high technology is utilized. The ways of doing business change and companies aim to produce high value-added products and services.

2.2.1 The Building Blocks of Industry 4.0

The building blocks of Industry 4.0 are generally considered to be Cyber-Physical Systems (CPS), Big Data, Internet of Things (IoT) and Smart Factories.

2.2.1.1 Cyber-Physical Systems

Today, the physical world and cyberspace are evaluated together. While the basis of cyberspace is based on the physical world, the boundaries of the physical world expand with cyberspace. Cyber-Physical Systems (CPS) that combine these two areas consist of a network created by objects and systems that communicate with each other via the Internet and an assigned Internet address. It also consists of a virtual environment that emerges within the computer environment simulation of objects and behaviors in the real world (Soylu, 2018: 46). CPS components are shown in Fig. 2.2. These components are Radio Frequency Identification (RFID) and IoT, Internet, embedded systems, wireless sensor networks (WSN), mobile networks and satellite networks.

CPS are systems where physical objects and computing resources are tightly integrated and have a continuous coordination between each other (Krämer, 2014: 1). Computers and networks often observe and control physical processes through feedback loops. While the physical processes affect the calculations, the calculations also affect the physical processes (Lee, 2008:363). CPS consists of hardware, processors and communication technologies supported by softwares. CPS provides machine-to-machine

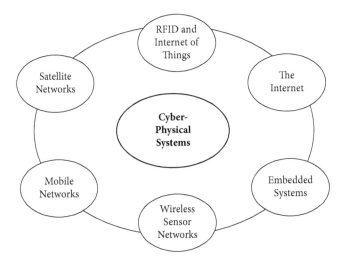

Fig. 2.2: Cyber-Physical Systems Components (Abosaq et al., 2016: 32)

(M2M) communication, autonomy as well as human machine cooperation, distributed intelligence and adaptive control, analysis and prediction of Big Data (Tu et al., 2018a: 100). CPS independently trigger actions, control each other and exchange information in real time on virtual networks (Stock and Seliger, 2016: 537; Maslarić et al., 2016: 511; European Parliament, 2016: 22).

2.2.1.2 Big Data

Big Data refers to complex data sets that are difficult to analyze with traditional data processing techniques and/or those structured and unstructured sets where algorithms do not work (Kang et al., 2016: 119; Sayki, 2016: 1). In a study by the Foundation for Industrial and Technical Research at the Norwegian Institute of Technology (SINTEF), it was found that 90 % of all data in the world has been generated in the last two years and the amount of data increases day by day. As the amount of memory needed is more than the amount of available memory, there is too much data to be analyzed at once. However, with the development of cloud computing services, it is easier to store the data and reach it when

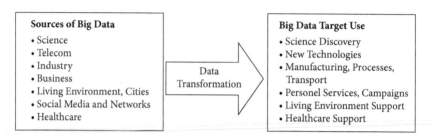

Fig. 2.3: Big Data Sources and Target Use (Demchenko et al., 2014: 105)

needed. The combination of the enormous dimensions of Big Data and the complexity of the analyses required to benefit from it leads to the development of a new class of technologies and tools to manage them (Doğan and Arslantekin, 2016: 22; Zhou et al., 2015: 2150). In the evaluation of the available data, software that can perform advanced analyses is used.

There are difficulties in extracting meaningful data quickly and easily from data sources. However, obtaining significant information and competitive advantages from large amounts of data has become increasingly important for businesses in international trade (Zakir et al., 2015: 81). Fig. 2.3 shows the major data sources and usage targets in general.

It is possible to obtain information on supply chain processes by benefiting from Big Data. Consumer behaviors are evaluated by conducting analysis that takes demographic characteristics, in particular, into consideration. With the data obtained, demand forecasts can be made for producers, suppliers and distribution channels, and allowing for production, stock and transportation processes to be planned.

2.2.1.3 Internet of Things

The Internet of Things (IoT) defines intelligent objects that can interact with each other through wireless and wired connections and unique addressing schemes, create new applications/services and collaborate with other objects to achieve common goals (Vermesan et al., 2013: 8). IoT is a network of sensors, software and physical objects that come together and are described in general as follows (Georgescu and Popescul, 2015):

- Control Devices: Smartphones, tablets and other smart devices that can control all objects.
- Cloud Service: Provides storage and access control between the object and its controller.
- Global Network: Many things are connected to the Internet in the global network, except for power transmission networks or secret government systems.
- Local Area Network: Consists of a controller area network, local networks in the home, and so on.
- Things: Things that can be remotely controlled by cloud services, the global network, and local area network.

The things are connected to the Internet and to each other through minimum human intervention, via WSN to meet their needs. As IoT technology is used by more companies, applications are expected to seamlessly integrate information and physical object flow and turn companies into a real-time business (Tu et al., 2018b: 66). IoT technology focuses on the entire ecosystem, including a supply chain, rather than a single company. Thus, all parties improve their processes resulting in maximizing benefits for end consumers (Szozda, 2017: 404). Furthermore, it provides solutions for companies to make strategic decisions and increase productivity.

ARKAS

Arkas Holding moved all servers and data centers to the cloud with Microsoft solutions. Lucien Arkas, Chairman of the Holding's Board, gave an example when sharing the impressions of the group after all of the servers and data centers were moved to the cloud: "The workers used to check the containers discharged from the ship one by one. They used to look at possible splitting and crushing in the container. That was the duty of 20–30 workers. Now, our security cameras are smart with chip support. Now the smallest crushing in the containers is captured by the cameras alerting us and then necessary interventions are carried out". Bernard Arkas, Vice Chairman of the Board of Directors, said, "one of the most important benefits is to be able to track the systems and organizational software packages of all our companies from one end to the other. It makes a big contribution to savings rates. Now we are free from the trouble of obtaining a server. We will be able to better manage our capacity. Using new technologies with more appropriate budgets and investing in systems and technologies according to new or changing business plans, will provide a big cost advantage without making system investments that may create unused capacity. This transition has brought us many benefits besides international standardization and discipline" (Munyar, 2019; DHA, 2019).

2.2.1.4 Smart Factories

A Smart Factory is a factory environment in which machinery and equipment can develop, maintain and control their own operation processes through automation and self-optimization. A U-Factory (ubiquitous factory), the factory of things, the factory in a real time-frame and the intelligent factory of the future are also used synonymously with the term "Smart Factory" (Hozdić, 2015: 31; Yoon et al., 2012: 2178; Lucke et al., 2008: 115; Zuehlke, 2010: 131; Hameed et al., 2011: 327). In such factories, operations are performed between four walls, but can also be connected to similar production systems in the global network and to a digital supply network within a broader framework (Deloitte, 2017: 5). Smart Factories carry out flexible and dynamic operations, adapt to complex production processes and produce solutions (Lee, 2015: 231). The machines in the production lines of smart factories evaluate their own performance, diagnose potential defective components, predict the risks that may cause failure, and share information in cyberspace to prevent these risks.

TOFAŞ

In the TOFAŞ plant, there are many assembly stations, thousands of operators and many industrial blockers used by these operators. TOFAŞ management have established a technology partnership with trexDCAS to collect data from these industrial blockers. With the "Smart Station 4.0 Project", which was carried out by TOFAŞ and trexDCAS, the necessary equipment and software were added to the blockers in order to transform them into smart stations used in the factory. In this way, it became possible to follow the desired torque and number of tightening. In addition, there are various projects carried out by TOFAŞ on the digitalization process. One of them is the "Collaborative Robot (Cobot)" project. Advanced safety equipment and software are integrated into medium load capacity robots. The robots, who are taught what to do when they interact with people, work together with people. For the vehicles, there are also studies conducted under their "Connectivity Project". With this project, vehicles use their own information for its own treatment and transfer that information to the driver or service (Ünlü, 2017; Tanık, 2018; Tokçalar et al., 2016: 40).

2.2.2 The Effects of Industry 4.0 on Competition

The effects of Industry 4.0 on the competition between companies come to the forefront due to a decrease in maintenance, quality and stocking costs;

shorter times that machines do not work, shorter durations for reaching the market and increased efficiency of technical personnel. As a result of these gains, it is aimed to achieve a total productivity increase. This productivity increase, which will be captured by the industries that direct the economy of a country, increases competitiveness in national and global markets. For this reason, companies focus on technology and innovations to maximize the opportunities of the new industrial revolution and integrate the applications required by Industry 4.0 into their business processes. In this regard, the main focal points of competition for various industries are as follows (Farahani et al., 2015: 31):

- In the consumer products industry, the focus is on managing the digital customer better and providing higher stock and demand visibility in the global network, and responding to changing customer needs more quickly and flexibly.
- In the high-tech products industry, it is aimed to standardize and automate processes by focusing on the simplification of the supply chain and an integration of supply chain platforms supported by big and intelligent data.
- In the machinery industry, the focus is primarily on using innovative technologies to sustain new business models.
- In the automotive industry, the main focus of the original equipment manufacturers and suppliers is to revise and develop processes as well as information and communication technology systems to cope with the unique challenges of globalization and supply chain.

2.3 Cyber Networks in Supply Chain and Logistics Management Processes

Large quantities of operational data collected from cyberspace are presented to the upstream and downstream supply chain through the systems and software used in business processes, and thus impacts beyond the boundaries of the facility. Consequently, the affix cyber is integrated with the concepts of supply chain and logistics management and it enables these concepts to be redefined to emphasize the information and communication technologies aspects.

The cyber supply chain includes the key actors (end users, policy makers, purchasing specialists, system integrators, network providers and software/hardware suppliers) involved in and/or utilizing the cyber infrastructure and their institutional and process interactions in order to plan, create, manage, protect and defend the cyber infrastructure (Boyson et al., 2009: 5). Production processes, supplier relations and customer expectations have changed with digitalization and this change requires logistics service providers to adapt to conditions in order to meet customer needs, optimize business processes and reduce costs (Maslarić et al., 2016: 512). Cyber logistics includes logistics services such as transportation, storage of goods and so on in cyberspace. It covers the information and communication technology processes for planning logistics services to increase online accessibility to information and the management of subsequent complementary processes (Chan et al., 2010: 288). It is aimed to develop more flexible and intelligent logistics solutions supported by IoT and CPS (Barretoa et al., 2017: 1248).

The data is obtained in real time and safely via cyber networks used in supply chain and logistics management. In addition, all processes related to the demand, production and resource planning, purchasing, stock and warehouse management, planning processes for transportation and distribution of goods, which are carried out through these data, are managed more efficiently.

2.3.1 Enterprise Resource Planning and Configuring the Future

Various kinds of software have been developed and used over time to manage business processes related to demand, production and resource planning in line with the needs of companies. In the 1960s, the Material Requirements Planning (MRP) system was created in order to provide the materials needed for production by processing the data of stock, order, sales forecasts, and so on (Islam et al., 2013: 13). MRP plans the material resource, but other resources such as labor and capital need to be planned as well. In the 1980s, the Production Resource Planning (MRP II) system was introduced with the necessity of planning all the resources of the producer (Tanna and Vyas, 2017: 3). In the 1990s, Enterprise Resource Planning (ERP) systems started to be used in order to coordinate the activities of companies in different geographical regions, to integrate with the business environment, and to require management systems for business processes in the manufacturing sectors as well as in the service sectors (Rashid et al., 2002: 2).

According to the the Association for Operations Management (APICS), ERP is a structure created to plan all resources of a business, ranging from strategic planning to implementation. Process connections are automated with information and communication technology tools, software, information is exchanged between functional areas and business processes are carried out effectively. There are two basic expectations from the system. The first is to reduce costs by increasing productivity through computerized automation. The second is the development of the decision-making process, providing accurate and timely information across the company (Poston and Grabski, 2001: 271). ERP systems also support data organization for decision-making and analysis and are generally planned as modules that support functional areas such as finance, marketing, human resources, operations, purchasing and logistics. Real-time data sharing is enabled by using a common database among modules (APICS, 2011: 48). For real-time data exchange in business processes, ERP software must be integrated with Manufacturing Execution Systems (MES). In this respect, a Smart Factory is a CPS that integrates information systems, such as ERP and MES, with physical assets like machinery, conveyors, and products for flexible and agile production (Wanga et al., 2016: 159). In order to prevent the production process from being interrupted, there is a need to foresee machine failures and to predict the root cause of the problems. Smart factories of the future will allow real-time data exchange, having these prediction capabilities (Haddaraab and Elragala, 2015: 723).

Due to the developments in cyberspace, ERP systems are also improved and ERP is used for organizing, directing and controlling all processes. With the IoT, intelligent machines connect, communicate and perform real-time calculations. Machines can request periodic maintenance or communicate with the manufacturer to avoid the risk of being disabled due to the need for a new part. Wireless, machine-to-machine systems and the location and status of everything required at the factory are known at each stage and these systems can be connected to the ERP (Weyrich et al., 2014: 21). In this way, resource planning and production optimization can be conducted. Information is gathered centrally to enable employees, partners, suppliers and customers to access online, in a two-way manner, and to be aware of the process. Thanks to cloud computing, information is managed wirelessly without needing physical storage.

2.3.2 Digital Purchasing

Purchasing is a business function but it is also an action for the company's purchase of products and services. In order to add value to the business, purchasing carries out various activities such as supplier market research, supplier identification and selection, negotiation and contract process regulation, purchasing, supplier measurement and improvement, and purchasing systems development (Monczka et al., 2009: 8). The purchasing process consists of defining the need, selecting suppliers, negotiating and contract agreement, ordering, order follow up, control, and evaluation sub-processes (Gottge and Menzel, 2017: 36). One of the most important factors in the success of the companies in their activities is the right organization of the purchasing process from a strategic point of view. Rising prices and inflation, the need to control stock investments more effectively, the recognition of the importance of purchasing costs for profitability, loss in production and the increasing scarcity of some important materials are the main factors leading to the development of a strategic perspective (Quayle, 2006: 27). The use of information and communication technologies supports the sustainability of the strategic purchasing process and the limitation of uncontrolled purchases.

Purchasing can take many forms in cyberspace. The most common are the platforms that bring businesses and suppliers together who buy materials and/or products on the Internet. Suppliers load their capacity information to this platform. Purchasing companies also apply to the system to use these available capacities to meet their needs. Suppliers may load their own prices to the system, and also receive price offers from businesses that will purchase. The supplier checks the realism of the offered price and expected delivery time, and procures the purchase when the conditions are suitable. In addition, companies can allow suppliers to gain access to their databases. Suppliers define their products with stock codes and enter information about the product visuals and features. In this way, the supplier follows the orders from the company itself (Karaca and Demirtaş,

2010: 53). Furthermore, through the e-procurement method, information can be exchanged transparently with suppliers, and goods and services can be purchased (Nawi et al., 2017: 211). Moreover, by using algorithms and mathematical methods, the effect of the human factor can be minimized, and the purchasing process can be carried out in a more automated way in real time, allowing them to see the capacity fill rates, the type of material needed, the types and properties of the materials, in order for them to make pricing and price offers (Deloitte, 2016: 10). The most advanced point of the transformation in purchasing is the ability to use the IoT and CPS to supply the orders of parts that are needed for the repair and maintenance of the machines and tools used in the production of products or services. For example, a ship in the ocean can inform its manufacturer, or the authorized agent at the port where she will visit, about the part or parts she needs without using any human factor. All these developments related to purchasing provide time and cost savings to the company.

2.3.3 Smart Warehouses

The warehouse is the interim point that plays a critical role in the realization of a whole series of activities in the supply chain, from the raw material stage of the products to the production environment and from there to the distribution of consumption centers (Koster and Smidts, 2013: 1230; Kiefer and Novack, 1999: 19). The warehouse is a dynamic and strategic center used to create a competitive advantage by contributing to a professional perspective and systematic work, expertise, qualified workforce, shortening delivery times and reducing customer order times (Hompel and Schmidt, 2007: 3). Applications that are supported by information and communication technologies allow more devices to connect and communicate in cyberspace. This situation also affects the warehouse business processes in terms of real-time monitoring and management. Smart warehouses that carry out order collection, distribution, delivery and record-keeping processes without using paper, in an automated and unmanned way, are constructed (Liu et al., 2018: 1). Such fully automated warehouses require automated storage systems and robots, as well as custom-made equipment. Each equipment can be organized automatically and operations are coordinated with the principle of multiple agent systems. In addition, the system interacts with all equipment in the distribution process (EMANS, 2018). In smart warehouses, stock items communicate with machines, which also connect and communicate with other machines. Business processes are carried out in an unmanned, real-time, reliable and cost-effective way by ERP devices and warehouse management systems.

By using complex algorithms and a wi-fi connection, order picking and packing operations in the warehouse are conducted automatically by robot arms and/or stackers without needing manpower. The pallet required by the operator is defined and determined by the system. The forklift truck automatically knows which pallet to choose and moves from the shelf to the area of that pallet. The loading line or production line equipment reports to the forklift where to place the pallet. When the forklift completes the delivery, the system automatically updates its stock records by real-time inventory software. The internal transport made by the forklift is also carried out by other stackers, conveyors and even unmanned vehicles.

There is also Radio Frequency Identification (RFID) technology to be mentioned, which is used as a warehouse management system in order to carry out the warehouse workflow efficiently. The digital transformation in industry has revealed the true potential of RFID. IoT-based warehouse management systems use RFID and WSN to monitor raw materials, materials and products (Lee et al., 2018: 2756). The components of RFID, a type of automatic identification system used to reduce time and labor costs and to increase real-time data accuracy, are shown in Fig. 2.4. The system is a technology that uses radio waves for communication between a tag containing a microchip connected to an antenna and a reading

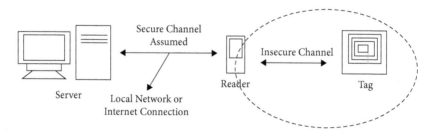

Fig. 2.4: Radio Frequency Identification System (Liao and Hsiao, 2014: 141)

device to read or encode data on the label. It enables a company to identify individual products and components and to track them from the point of production to the point of sale in the supply chain (Liao and Hsiao, 2014: 136).

RFID tags help components, products, machines, and even the workforce to work together more efficiently. This system is used in production environments as well as warehouses due to its benefits. The products tell the machines what is needed at a certain point. After each step in the process, the machine updates the product status by writing on the label and then sends the product to the next step. The tag's data can be accessed at any time for real-time updates and analysis, and the security mechanisms on the label's chip are used to prove originality, quality and security. Since labels can be accessed even if the product is in the package, options for advanced stage customizations, such as language or other country-specific settings, are created after leaving the factory.

2.3.4 Transport and Distribution Management in the Digital Age

The impact of digitalization on business processes and of the increasing demand for personalized special products and services are also reflected in transport and distribution services as in other stages of the supply chain. Operation processes in transportation have been changing with automation and digitalization. With RFID-WSN devices embedded in the delivered products, the cargo can be monitored in real-time, and data on the transport conditions are collected, stored and transmitted (Mejjaouli and Babiceanu, 2018: 69). The data of the transport vehicles can be accessed from the center momentarily, the transport process is monitored transparently, the waiting duration is shortened, and the autonomous or artificial intelligence vehicles reduce transportation and distribution costs. In fact, many work processes will be carried out in next generation vehicles that will have high-tech equipment. Work processes such as order taking, route planning, and moving stock that are currently conducted at the centers of logistics companies will be transferred to vehicles and vehicles will be transformed into mobile offices. In practice, the main developments that have occurred and/or are expected to occur in all modes of transport for the realization of these benefits are as follows:

- In maritime transportation, a large amount of data is recorded at sea and analyzed with access from land (Fruth and Teuteberg, 2017: 33). With CPS, the movements of equipment in the ports are programmed and monitored (Hribernik, 2016: 11). With new bridge functions, there is a transformation towards unmanned ships (Rødseth, 2016).
- In road transportation, autonomous and driverless transport systems will have significant developments in human and freight transportation in the future and research continues on these areas. Companies such as Google, Tesla, Uber and Waymo have been working on autonomous road transport vehicles. State-of-the-art technology road vehicles are equipped with multiple radio interfaces, radar, camera and other sensor devices so that they can make comprehensive calculations and communicate among vehicles and from vehicle to road infrastructure (Besselink et al., 2016: 1129). In the long term, autonomous trucks will communicate directly with terminals and other means of transport to ensure that loads are in the right place at the right time (D'inca et al., 2016: 6).
- In air transportation, CPS makes aircrafts more intelligent, and future air traffic management systems are designed as CPS to provide the necessary capacity, efficiency, safety and security system performance (Valdes and Comendador, 2018: 228).
- In rail transportation, CPS is used to observe, plan, control and manage train and rail equipment to meet operational objectives without compromising safety requirements (Zhang, 2013: 270). The wagons and containers connected to each other will provide their own workflow updates in terms of state, location and routes (D'inca et al., 2016: 6).

Additionally, ongoing research and development (R&D) studies on digitalization in transport and distribution operations are the indicators of greater developments in logistics services in the future.

2.4 Cyber Security in Supply Chain and Logistics Management

Data is the vital lifeblood of modern supply chains that are more collaborative, connected, and increasingly globally organized, and for this reason ensuring security of these chains is critical to healthy business operations (Sangster, 2015: 30). However, the use of information and communication technologies in business processes also brings new security problems. One of the risks of supply chain that threatens data security are cyber attacks with destructive, disruptive and unpredictable effects.

The risk of cyber attacks are the security vulnerabilities arising from the combination of threat, fragility and impact that arise from the source moving in the digital environment with a certain intent, as shown in Fig. 2.5 (Waalewijn, 2014: 24). A cyber attack disrupts the confidentiality, integrity and appropriateness of activities and/or the environment and undermines the success of economic and social objectives (OECD, 2015:30). Moreover, data such as transport programs, security system plans, personnel information from database, are used to determine the method and goal of the attack (Urciuoli et al., 2013: 61).

Although the concept of the cyber attack risk is relatively new, it has existed since the first use of the computer (Davis and Sullivan, 2017: 16). As more and more devices and control systems are connected online, there is a greater vulnerability to risks that may affect the operation of physical assets (MARSH, 2015: 2). Cyber attacks are carried out by organized crime syndicates, radical organizations, rebels, terrorists, foreign intelligence organizations, political activists, hackers, dissatisfied or opponent employees and researchers in various fields (Wamala, 2012: 16; Cambridge

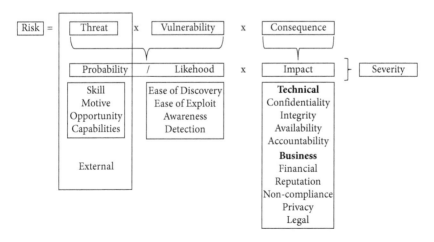

Fig. 2.5: Risk Calculation (DNV GL, 2018:7)

Centre for Risk Studies, 2017: 4; Bieda and Halawi, 2015: 38). The goal of cyber attacks are always the weakest point and this weakest point has been increasingly coming out of the supply chain. In cyber attacks, the easiest way for a company's systems and data is sought (KPMG, 2018: 4). Therefore, the cyber security of any company in a supply chain is as strong as the weakest member of the chain and generally, the weakest cyber security arrangements in the supply chain are present in small companies (CERT-UK, 2015: 3; Urciuoli, 2015: 16). With the attacks, data is deleted, changed, copied and system entries are prevented, and the functionality of the supply chain is damaged by disrupting the system's working order (Warren and Hutchinson, 2000: 712).

What is important in operations is not only the security of the company itself, but also the cyber security of the suppliers (Burnson, 2016: 49). In supplier relations, it is important to know who the suppliers and the supplier of their suppliers are, and how suppliers and partners manage their own cyber supply chain risks for products and services acquired from them (NIST, 2015: 2). Supplier cyber security competencies are identified by studies, such as surveys, as shown in Fig. 2.6. Furthermore, items are included in the contract for data security, the processes for data violations are developed and implemented in cooperation with suppliers, and regular information assurance activities are carried out to determine the critical pathways (CERT-UK, 2015: 9).

It should not be forgotten that there are technologies to help increase cyber security but no supply chain will ever be 100 percent secure (PwC, 2011: 9). The provision of cyber security in companies depends on information and communication technology security (information systems) through carrying out security tests and taking protective security measures on physical security (building, office, facility) by inventory security and access control, and operational technology security (plant automation) through monitoring in real-time operations including production, transportation and handling (Burns, 2016: 276).

Fig. 2.6: An Example of a Questionnaire to Evaluate Suppliers' Safety Practices and Standards (Pal and Alam, 2017: 663)

XYZ Company
Supplier Cyber Security Qualification Assurance Form

Document Number
Registration Date
Evaluation Date
Revision Date

Questions	**Descriptions**
Whether design process is documented?	
Is the design process of software or hardware product repeatable?	
How does vendor deal the existing and emerging vulnerabilities? How is vendor capable to address new vulnerabilities?	
What kind of standards is used by vendor to manage and monitor manufacturing, assembling, and testing processes?	
How is code quality tested?	
What techniques, procedures and approaches are used for protection and detection of malware?	
How "tamper proofing" of products is done? What are the methods for closing the backdoors?	
Whether all process are documented properly and audition is conducted as per standards?	
What kind of access controls is in place?	
How is customer's data protected?	
What is the encryption mechanism?	
How long is the retention period for data?	
What is the policy for data destruction when the partnership is dissolved?	
Whether background checks are performed for employees? If yes then how frequently?	
What kind of security practices is followed?	
Whether there is proper cyber security check list for upstream and downstream suppliers? How is adherence to check list?	
Whether proper security checks are performed for the distribution process?	
What are the selection criteria for selecting the distribution channels?	
What is the mechanism for disposing of the counterfeit component?	
How does vendor ensure the security of the process, product, service etc. throughout the product life cycle?	

References

Abosaq, N. H., Alandjani, G. and Pervez, S. (2016). "IOT Services Impact as a Driving Force on Future Technologiesby Addressing Missing Dots". *International Journal of Internet of Things and Web Services*, (1): 31–37.

Almagor, R. C. (2011). "Internet History". *International Journal of Technoethics*, 2(2): 45–64.

APICS. (2011). *Operations Management Body of Knowledge Framework*. Chicago: APICS The Association for Operations Management.

Barretoa, L., Amarala, A. and Pereiraa, T. (2017). "Industry 4.0 Implications in Logistics: An Overview". *Procedia Manufacturing*, (13): 1245–1252.

Besselink, B., Turri, V., van de Hoef, S. H., Liang, K.-Y., Alam, A., Martensson, J. and Johansson, K. H. (2016). "Cyber-physical Control of Road Freight Transport". *IEEE Special Issue Industrial Cyber-Physical Systems*, 104(5): 1128–1141.

Bieda, D. and Halawi, L. (2015). "Cyberspace Avenue for Terrorism". *Issues in Information Systems*, 16(3): 33–42.

Boulila, N. (2019). *Cyber-Physical Systems and Industry 4.0: Properties, Structure, Communication, and Behavior*. Germany: Siemens.

Boyson, S., Corsi, T. and Rossman, H. (2009). Building A Cyber Supply Chain Assurance Reference Model. A collaborative research project between SAIC and the Supply Chain Management Center (SCMC), Robert H. Smith School of Business. Accessed: 15.04.2019, http://citeseerx.ist.psu.edu/viewdoc/download?doi=10.1.1.363.1516&rep=rep1&type=pdf.

Burns, G. M. (2016). *Logistics and Transportation Security a Strategic Tactical and Operational Guide to Resilience*. Boca Raton: CRC Press.

Burnson, P. (August 2016). Ocean Cargo Roundtable: What's in Store for 2017?. Accessed: 16.11.2019, https://www.logisticsmgmt.com/article/ocean_cargo_roundtable_whats_in_store_for_2017.

Cambridge Centre for Risk Studies. (2017). *Cyber Terrorism: Assessment of the Threat to Insurance (November 2017)*. United Kingdom: University of Cambridge Judge Business School.

CERT-UK. (2015). *Cyber-Security Risks in the Supply Chain*. United Kingdom: CERT-UK.

Chan, Y., Fredouët, C. H., Boukachour, J., Lo, H. P., Chiang, C. C., Moeeni, F., Dey, M. M. and Toh, A. K. (2010). "Cyber Transportation Logistics: Architecting a Global ValueChain for Services". In Lou, Z. (Ed.), *Service Science and Logistics Informatics: Innovative Perspectives*, (272–297). New York: IGI Global.

Davis, J. and Sullivan, J. (March–April 2017). "Supply Chain Risk— What Is It?". *Defense AT&L*, XLVI(2): 15–18.

Deloitte. (2016). Industry 4.0 and distribution centers transforming distribution operations through innovation. Accessed: 15.04.2019, https://www2.deloitte.com/content/dam/insights/us/articles/3294_ industry-4-0-distribution-centers/DUP_Industry-4-0-distribution- centers.pdf.

Deloitte. (2017). The smart factory Responsive, adaptive, connected manufacturing A Deloitte series on Industry 4.0, digital manufacturing enterprises, and digital supply networks. Accessed: 14.04.2019, file https://www2.deloitte.com/content/dam/insights/us/articles/4051_The- smart-factory/DUP_The-smart-factory.pdf.

Demchenko, Y., Laat, C. and Membrey, P. (2014). "Defining Architecture Components of the Big Data Ecosystem". In Smari W.W. (Ed.), Fox, G. C. (Ed.) and Nygård, M. (Ed.), *Proceedings of the International Conference on Collaboration Technologies and Systems*, (104–112). Minneapolis, MN: IEEE.

D'inca, J. Reinhold, T. and Sitte, T. (2016). Taking Rail Virtual Through Digital Industry. Accessed: 16.11.2019, https://www.oliverwyman .com/content/dam/oliver-wyman/v2/publications/2016/apr/FINALTran sportLogisticsJournal2016BookFull-web.pdf

DNV GL. (June 2018). Probability and Uncertainty Whitepaper Some Things Are More Uncertain Than Others. Accessed: 16.04.2019, https://www.dnvgl.com/Images/Cyber-security-whitepaper-Probability- and-uncertainty_tcm8-137894.pdf.

DHA. (2019). "Arkas rotasını buluta çevirdi". Hürriyet, Accessed: 23.05.2019, http://www.hurriyet.com.tr/teknoloji/ arkas-rotasini-buluta-cevirdi-41124004.

Doğan, K. and Arslantekin, S. (2016). "Büyük Veri: Önemi, Yapısı ve Günümüzdeki Durum". *Ankara University Journal of the Faculty of Languages and History-Geography*, (56)1: 15–36.

EMANS. (July 2018). The Road to Automated and Intelligent Warehouse. Accessed: 23.05.2019, https://www.anasoft.com/emans/en/home/news-blog/blog/The-Road-to-Automated-and-Intelligent-Warehouse.

Eureopean Commission. (2017). Digital Transformation Monitor Germany: Industrie 4.0. Accessed: 15.02.2019, https://ec.europa.eu/growth/tools-databases/dem/monitor/sites/default/files/DTM_Industrie%204.0.pdf.

European Parliment. (February 2016). Industry 4.0 (PE 570.007). Accessed: 11.04.2019, http://www.europarl.europa.eu/RegData/etudes/STUD/2016/570007/IPOL_STU(2016)570007_EN.pdf.

Farahani, P., Meier, C. and Wilke, J. (2015). "Digital Supply Chain Management 2020 Vision". *360°—the Business Transformation Journal*, (13): 20–33.

Felea, M. and Albastroiu, I. (2013). "Defining the Concept of Supply Chain Management and Its Relevance to Romanian Academics and Practitioners". *Amfiteatru Economic*, 15(33): 74–88.

Frenzel, C. W. and Frenzel, J. C. (2004). *Management of Information Technology*. Boston, MA: Cengage Learning, Inc.

Fruth, M. and Teuteberg, F. (2017). "Digitization in Maritime Logistics – What Is There and What Is Missing?". *Cogent Business and Management*, (4): 1–41.

Georgescu, M. and Popescul, D. (2015). "Security, Privacy and Trust in Internet of Things: A Straight Road?" [Proceeding]. The 25th IBIMA conference on Innovation Vision 2020: from Regional Development Sustainability to Global Economic Growth (IBIMA 2015), 7–8 May 2015, Amsterdam, Netherlands.

Gottge, S. and Menzel T. (2017). Purchasing 4.0: An Exploratory Multiple Case Study on the Purchasing Process Reshaped by Industry 4.0 in the Automotive Industry. Linnaeus University, Master Thesis, Sweden.

Haddaraab, M. and Elragala, A. (2015). "The Readiness of ERP Systems for the Factory of the Future". *Procedia Computer Science*, (64): 721–728.

Hameed, B., Durr, F. and Rothermel, K. (2011). RFID based Complex Event Processing in a Smart Real — Time Factory, Expert discussion: Distributed Systems in Smart Spaces.

Hompel, M. and Schmidt, T. (2007). *Warehouse Management*. Berlin: Springer.

Hozdić, E. (2015). "Smart Factory for Industry 4.0: A Review". *International Journal of Modern Manufacturing Technologies*, 7(1): 28–35.

Hribernik, K. (2016). Industry 4.0 in The Maritime Sector Potentials and Challenges, BIBA - Bremer Institut für Produktion und Logistik GmbH, Tokio.

Islam, Md. S., Rahman, Md. M., Saha, R. K. and Saifuddoha A. Md. (2013). "Development of Material Requirements Planning (MRP) Software with C Language". *Global Journal of Computer Science and Technology Software & Data Engineering*, 13(3): 12–22.

Janvier-James, A. M.(2012). "A New Introduction to Supply Chains and Supply Chain Management: Definitions and Theories Perspective". International Business Research, 5(1): 194–207.

Kang, H. S., Lee, J. Y., Choi, S., Kim, H., Park, J. H., Son, J. Y., Kim, B. H. and Noh, S. D. (2016). "Smart Manufacturing: Past Research, Present Findings, and Future Directions". *International Journal of Precision Engineering and Manufacturing-Green Technology*, 3(1): 111–128.

Karaca, Y. and Demirtaş M. (2010). "E-Tedarik Sistemlerinin İşletme Performansina Etkisi ve Dengeli Skor Kart ile Performans Ölçümü". *ZKU Journal of Social Sciences*, 6(11): 47–62.

Kiefer, A. W. and Novack, R. A. (1999). "An Empirical Analysis of Warehouse Measurement Systems in the Context of Supply Chain Implementation". *Transportation Journal*, 38(3): 18–26.

Koster, M. B. M. and Smidts, A. (2013): "Organizing Warehouse Management". *International Journal of Operations & Production Management*, 33(9): 1230–1256.

KPMG. (2018). Digital Supply Chain — the Hype and the Risks (N16151LOBS). Australia: KPMG.

Krämer, B. J. (2014). "Evolution of Cyber-Physical Systems: A Brief Review". In Suh, S. C. (Ed.),**Tanik**, U. J. (Ed.),**Carbone**, J. N.

(Ed.) and **Eroglu**, A. (Ed.). *Applied Cyber-Physical Systems*, (1). New York: Springer.

Lee, C. K. M., Lv, Y., Ng, K. K. H., Ho, W. and Choy K. L. (2018). "Design and Application of Internet of Things-based Warehouse Management System for Smart Logistics". *International Journal of Production Research*, 56(8): 2753–2768.

Lee, E. A. (2008). "Cyber Physical Systems: Design Challenges" [Proceeding]. International Symposium on Object/Companent/Service-Oriented Real-Time Distributed Computing (ISORC). May 6, 2008, Washington, DC, USA.

Lee, J. (2015). "Smart Factory Systems". *Informatik—Spektrum*, 38(3): 230–235.

Lee, T. J. B., Cailliau, R. and Groff, J. F. (1992). "The World-Wide Web". *Computer Networks and ISDN Systems*, (25): 454–459.

Liao, Y. P. and Hsiao, C. M. (2014). "A Secure ECC-based RFID Authentication Scheme Integrated with ID-verifier Transfer Protocol". *Ad Hoc Networks*, (18): 133–146.

Liu, X., Cao J., Yang, Y. and Jiang S. (2018). "CPS-Based Smart Warehouse for Industry 4.0: A Survey of the Underlying Technologies". *Computers*, 7(13): 1–17.

Lucke, D., Constantinescu, C. and Westkämper, E. (2008). *Smart Factory—A Step Towards the Next Generation of Manufacturing, in Manufacturing Systems and Technologies for the New Frontier*. London: Springer, pp. 115–118.

MARSH. (2015). Cyber Risk in the Transportation Industry. London: MARSH and McLENNAN Companies.

Maslarić, M., Nikoličić. S and Mirčetić, D. (2016). "Logistics Response to the Industry 4.0: the Physical Internet". *Central European Journal of Engineering*, (6): 511–517.

Mejjaouli, S. and Babiceanu, R. F. (2018). "Cold Supply Chain Logistics: System Optimization for Real-time Rerouting Transportation Solutions". *Computers in Industry*, (95): 68–80.

Monczka, R. M., Handfield, R. B., Giunipero, L. C. and Patterson, J. L. (2009). *Purchasing and Supply Chain Management*. United States of Americe: South Western Cengage Learning.

Munyar, V. (2019). "Bulut, teknolojiye demokrasi getirdi 112 bin iş kapısı açacak". Hürriyet. Accessed: 23.05.2019, http://www.hurriyet.com.tr/yazarlar/vahap-munyar/bulut-teknolojiye-demokrasi-getirdi-112-bin-is-kapisi-acacak-41172864.

NIST. (October 2015). Business Case for Cyber Supply Chain Risk Management. Accessed: 16.04.2019, https://csrc.nist.gov/CSRC/media/Projects/Supply-Chain-Risk-Management/documents/briefings/Workshop-Brief-on-Cyber-SCRM-Business-Case.pdf.

Nawi, M. N. M., Deraman, R., Bamgbade, J. A., Zulhumadi, F. and Riazi, S. R. M. (2017). "E-Procurement in Malaysian Construction Industry: Benefits and Challenges in Implementation". *International Journal Supply Chain Management*, 6(1): 209–213.

OECD. (2015). *Digital Security Risk Management for Economic and Social Prosperity OECD Recommendation and Companion Document*. Paris: OECD.

Öztürk, D. (2017). "Technological Transformation of Manufacturing by Smart Factory Vision: Industry 4.0". *International Journal of Development Research*, 7(11): 17371–17382.

Pal, O. and Alam, B. (2017). "Cyber Security Risks and Challenges in Supply Chain". *International Journal of Advanced Research in Computer Science*, 8(5): 662–666.

Poston, R., and Grabski, S. (2001). "Financial Impacts of Enterprise Resource Planning Implementations". *International Journal of Accounting Information Systems*, 2(4): 271–294.

PwC. (June 2011): Transportation & Logistics 2030 Volume 4: Securing the supply chain. Accessed: 16.04.2019, https://www.pwc.com/gx/en/transportation-logistics/pdf/tl2030_vol.4_web.pdf.

Quayle, M. (2006). *Purchasing and Supply Chain Management: Strategies and Realities*. United States of America: Idea Group.

Rashid, M. A., Hossain, L. and Patrick, J. D. (2002). "The Evolution of ERP Systems: A Historical Perspective". In Rashid, M. A. (Ed.), Hossain, L. (Ed.) and Patrick, J. D. (Ed.), *Enterprise Resource Planning: Global Opportunities and Challenges*, (1–16). United States of America: Idea Group.

Rødseth, Ø. J. (2016). "Sustainable and Competitive Cyber-Shipping through Industry 4.0" [Proceeding]. Singapore Maritime Sustainability Forum 2016: Smart Maritime Solutions and Overcoming Challenges, 19 April 2016, Singapore.

Sangster, S. (October 2015). "Scrutinizing Content for Greater Supply Chain Security". MHL News, 28–31.

Sayki, K. T. (2016). Big Data: Understanding Big Data. Accessed: 11.04.2019, arXiv, 1601.04602, https://arxiv.org/ftp/arxiv/papers/1601/1601.04602.pdf.

SINTEF. Big Data-for better or worse. Accessed: 18.09.2018, https://www.sintef.no/en/latest-news/big-data-for-better-or-worse/.

Soylu, A. (2018). "Endüstri 4.0 ve Girişimcilikte Yeni Yaklaşımlar". *Pamukkale University Journal of Social Sciences Institute*, (32): 43–57.

Stock, T. and Seliger, G. (2016). "Opportunities of Sustainable Manufacturing in Industry 4.0". *Procdia CIRP*, (40): 536–541.

Szozda, N. (2017). "Industry 4.0 and Its Impact on The Functioning Of Supply Chains". *Scientific Journal of Logistics*, 13(4): 401–414.

Tanık, M. (2018). "trexDCAS: TOFAŞ'ta Akıllı İstasyon 4.0 projesi gerçekleştirdik". ST Endüstri Medya. Accessed: 23.05.2019, https://www.stendustri.com.tr/endustri-40-uygulamalari/trexdcas-tofasta-akilli-istasyon-40-projesi-gerceklestirdik-h94244.html.

Tanna, J. and Vyas, A. (2017). "Case Study on Manufacturing Resource Planning". *Claro: Journal of Engineering*, (37): 1–7.

Tokçalar, Ö. K., İlhan, R., Şimşek, Ö. and Durgun, İ. (2016). "Hibrit montaj sistemleri orta yük kapasiteli robotlarda insan-robot etkileşimi". In Dilibal, S. (Ed.), Şahin, E. (Ed.), Şahin, H. (Ed.), Kalkan, S. (Ed.) and Sarıel, S. (Ed.). *TÜRKİYE ROBOT BİLİM KONFERANSI ToRK 2016 Bildiriler Kitabı*, (40–43). İstanbul: Özlem Matbaacılık.

Tu, M., Lim, M. K. and Yang, M. F. (2018a). "IoT-based Cyber-physical System: A Framework and Evaluation". *Industrial Management & Data Systems*, 118(1): 96–125.

Tu, M., Lim, M. K. and Yang, M. F. (2018b). "IoT-based Production Logistics and Supply Chain System – Part 1 Modeling IoT-based Manufacturing Supply Chain". *Industrial Management & Data Systems*, 118(1): 65–95.

Urciuoli, L. (April 2015). "Cyber-Resilience: A Strategic Approach for Supply Chain Management". *Technology Innovation Management Review*, 5(4): 13–18.

Urciuoli, L., Männistö, T., Hintsa J. and Khan, T. (2013). "Supply Chain Cyber Security – Potential Threats". *Information & Security: An International Journal*, (29): 51–68.

Ünal, A. N. and Ergen, A. (2018). "Siber Uzayda Yeterince Güvenli Davranıyor muyuz? İstanbul İlinde Yürütülen Nicel Bir Araştırma". *Manisa Celal Bayar University Journal of Social Science*, 16(2): 191–216.

Ünal, A. N. and Saygılı, M. S. (2019). "Sanayi 4.0 Dönüşümünde 4 Boyutlu Baskı Teknolojisinin Yeri ve Tedarik Zinciri Yönetimine Etkileri". *Duzce University Journal of Science & Technology*, 7(2): 1–14.

Ünlü, E. D. (30.11.2017). "Endüstri 4.0'ı kendi dinamiklerimize göre yorumlamalıyız". Dünya. Accessed: 23.05.2019, https://www.dunya.com/kose-yazisi/endustri-40i-kendi-dinamiklerimize-gore-yorumlamaliyiz/392668.

Valdes, R. A. and Comendador, V. F. G. (2018). "Aviation 4.0: More Safety Through Automation and Digitization". *WIT Transactions on the Built Environment*, (174): 225–236.

Vermesan, O., Friess, P., Guillemin, P., Sundmaeker, H., Eisenhauer, M., Moessner, K., Le Gall, F. and Cousin, P. (2013). "Internet of Things Strategic Research and Innovation Agenda". In Vermesan, O. (Ed.) and Friess P. (Ed.). *Internet of Things—Converging Technologies for Smart Environments and Integrated Ecosystems*, (7–152). Denmark: River Publishers.

Waalewijn, D. (2014). *Cyber Security in the Supply Chain of Industrial Devices*. University of Twente, Master Thesis, Nederland.

Wamala, F. (2012). *The ITU National Cyber Security Strategy Guide*. Geneva: International Telecommunication Union.

Wanga, S., Wana, J., Zhang, D., Li, D. and Zhanga, C. (2016). "Towards Smart Factory for Industry 4.0: A Self-organized Multi-agent System with Big Data Based Feedback and Coordination". *Computer Networks*, (101): 158–168.

Warren, M. and Hutchinson, W. (2000). "Cyber Attacks Against Supply Chain Management Systems: A Short Note". *International Journal of Physical Distribution & Logistics Management*, 30(7/8): 710–718.

Weyrich, M., Schmidt, J-P. and Ebert, C. (2014). "Machine-to-Machine Communication". *IEEE SOFTWARE*, 31(4): 19–23.

World Economic Forum. (2012). New Models for Addressing Supply Chain and Transport Risk. Accessed: 08.03.2019, http://www3.weforum.org/docs/WEF_SCT_RRN_NewModelsA ddressingSu pplyChainTransportRisk_IndustryAgenda_2012.pdf.

Yarman, S. B. and Ünal, A. N. (2015). *Stratejik Karar Verme Boyutunda Bilgi Toplama/İşleme Amaçlı Karar Destek Sistemleri*. Ankara: Nobel.

Yoon, J. S., Shin, S. and Suh, S. H. (2012). "A Conceptual Framework for the Ubiquitous Factory". International Journal of Production Research, 50(8): 2174–2189.

Zakir, J., Seymour, T. and Berg, K. (2015). "Big Data Analytics". *Issues in Information Systems*, 16(2): 81–90.

Zhang, L. (2013). "Modeling Railway Cyber Physical Systems Based on AADL". In Cao, Y. (Ed.), Qin, S. (Ed.) and Grema, A. S. (Ed.), *Proceedings of the 19th INTERNATIONAL CONFERENCEON ON AUTOMATION AND COMPUTING Future Energy and Automation*, (270–275). London: IEEE.

Zhou, K., Liu, T. and Zhou, L. (2015). "Industry 4.0: Towards Future Industrial Opportunities and Challenges". In Tang, Z. (Ed.), Jiayi Du, J. (Ed.), Yin, S. (Ed.), He, L. (Ed), and Li, R. (Ed), *12th International Conference on Fuzzy Systems and Knowledge Discovery*, (2147–2152). Zhangjiajie, China.

Zuehlke, D. (2010). "Smart Factory—Towards a Factory—of—Things". *Annual Review in Control*, 34(1): 129–138.

Ahu Ergen

3. Smart Retailing in Cyberspace

Abstract: In highly competitive business environment, brands are trying to differentiate themselves from the competitors and try to offer superior experience to their customers. Production is easy today. However, the problem is how to differentiate and sell these products. Retailers are playing an important role at this point. Many retailers today use smart technologies such as eye-tracking, shopping path recording, shopping cards, shelf-talkers, in-store TV and face recognition to enhance the consumer shopping experiences. New innovations enhanced by technology are on the way to serve the retailers. This chapter covers changes in consumer and retailing environment, smart retailing tools and future of retailing.

Keywords: Smart Retailing, Smart Technologies, Marketing, Future of Retailing

3.1 Introduction

In today's dynamic competitive environment, retailing is changing with a huge speed. One of the major reasons of this change is the rapid technological developments. Many retailers have started to use smart technologies such as eye-tracking, shopping path recording, shopping cards, shelf-talkers, in-store TV and face recognition to enhance the consumer shopping experiences. Retailers try to follow changing consumer trends and adapt technological developments to their marketing strategies. Especially Generation Z and Generation Alpha members integrate technology to their lives more easily and use it efficiently. This will bring new smart technologies to the focus of retailers in the close future.

Consumers who were going out to shop as a social and entertaining activity are now searching prices on Internet and sharing pages with each other, following influencers for their brand choices. Some consumers are using the stores as showrooms, after seeing and touching the physical product, coming back home and start comparing prices from their smart devices. On the other side, consumers and brands face new challenges to prevent cyber attacks and data breaches.

Today, offering a holistic shopping experience to the customer is vital since today's customers are omnichannel both for attitude and behavior. Retailers today are trying to enhance the shopper experience with smart retailing tools to get valuable customer insights, save customers' time and make them feel independent. From their side, they can build positive image, build reputation, reduce their costs, gain new customers and keep the current ones. Also, such retailers may have the opportunity to be perceived more innovative and they can be imitated hardly by competitors (Savastano et al., 2019). Since number of technology-dependent consumers are increasing, in order to gain competitive advantage, companies need to be good at this adaptation. Internet of things, cloud technology, wearable technologies, use of big data, socially interactive dressing rooms, interactive mirrors and in-store apps are only some of the most popular technologies used in today's retailing space.

Bekoglu and Ergen (2016) group the technologies used in retailing under three categories which are smart technologies, biometric technologies and technologies developed for different dimensions of reality. While providing many advantages for retailers and consumers, these new technologies have also high switching and implementation costs.

3.2 Changes in Consumer and Retailing Environment

With the growth of companies such as Amazon and Alibaba, which sell through digital channels, retailing industry's dynamics started to change. Also many small-sized companies entered into e-trade market. Deloitte Research (2018) shows that the digital device (smart phones, in-store kiosks, laptops and desktops) effect on physical store sales was 14 % in 2013 and it reached 56 % in 2016. This rapid growth in three years time shows the increasing role of digital technologies in the physical store environment. Some consumers use the physical store just as a showroom and complete all the other buying decision steps on digital platforms.

One of the biggest problems that today's retailers face is pulling the consumers to the stores again. One way to solve this problem may be to bring the digital experience in the store. This can partly be achieved by new smart technologies such as physical showrooms for on-line sales, smart mirrors, augmented and virtual realities, checkout-free stores, face-based buying, voice-based and visual search, chatbots and drones (https://www.cmo.com). Irfan et al. (2019) state that customers today like to enjoy their shopping experience with convenience and maximum value for their money. In-store logistics and store image are also important indicators of customer's satisfaction today.

On the other side, for many retailers Generation Z is becoming an important target group while their consumption style is very different from the other generations. Priporas, Stylos and Fotiadis (2017: 374) conducted a qualitative research with 38 university students in UK and found out that smart technologies highly influence their consumption. Specifically, they require new devices and electronic processes to get more autonomy and faster transactions. They also want the technology to enable them to make more informed shopping decisions. For these young consumers, training is also an important topic, because they want to learn how to use these new smart retailing applications. Pantano et al. (2018: 95) state that the new technologies such as interactive displays, storefronts and signage used in retailing improves the consumers' shopping experience by entertaining and pleasing them. This brings more access points and elements able to engage more consumers.

Shankar et al. (2011: 31–32, 40) state that demographics, psychographics and behavioral characteristics of shoppers moderate the effects of the drivers on shopper behavior which effects innovations in shopper marketing. Most innovations that the manufacturers and the retailers use stem from smart technologies. These innovations reshape shopper behavior. The relationship between shopper behavior and shopper marketing innovations is bidirectional as seen in Fig. 3.1.

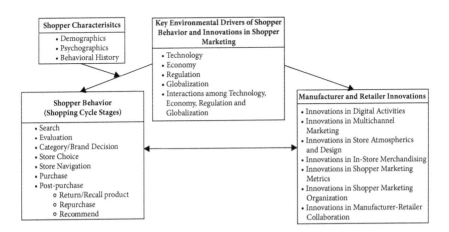

Fig. 3.1: A Framework for Analyzing Innovations in Shopper Marketing (Shankar et al., 2011: 31)

Still many companies don't focus on how, where and why the customer shops. Without understanding the daily lives of the customers, implementing technology in their retailing setting is not going to be meaningful. In most cases, the retailers are only trying to digitalize the physical experience of the customers (https://www.zdnet.com). In order to be successful, it is recommended that the retailers have to develop their retailing concepts around the lifestyles of their customers. The future of retailing depends on "tailing" the shoppers which follows the behavior models of the consumers in their daily activities (how and where they live, work and shop) and design their strategies accordingly (Fig. 3.2). So, shopping activity is going to be seen as a part of daily life routine. Innovative retailers personalize the products according to the needs and wants of the customers and begin to be closer with them (Turano, 2012).

Fig. 3.2: Tailing Pyramid (Turano, 2012)

Today's postmodern consumers like to be individual and reexplore themselves. Consumers prefer an interactivity, connectedness and creativity-based marketing communications in their consumption experiences and this is can be provided by smart marketing technologies (Vrontis et al., 2017: 273).

3.3 What Is Smart Retailing?

Smart retailing means integrating the traditional shopping methods with smart technologies. By using IoT technology, data is stored via communication between devices and computers. As a result, consumers enjoy more personalized, faster and smarter shopping experience. After the first use of scanner for a pack of chewing gum's sales in USA in 1974, many major technological innovations have revolutionized retailing (Priporas, Stylos and Fotiadis, 2017: 375). Today, scanner is a must in every retail store. Retailers are using RFID, mobile technology, TV network, holograms, and virtual reality in their retail marketing operations (Shankar et al., 2011: 34). So, retailing is also taking place in cyberspace, not only in the physical stores. Smart retailing is different from e-retailing which is mainly digital. It bridges digital and physical. Regarding space, core technology, nature of interactivity, experience and service provision there are differences between e-retailing and smart retailing (Tab. 3.1).

Smart retailing also refers to consumers' interactions with innovative technologies that aim to improve the shopping experience. It is facilitating consumer access to products, services and crucially, information (Vazquez et al., 2017: 425). Roy et al. (2017: 259) define smart retailing as *"an*

Tab. 3.1: Key Differences Between e-Retailing and Smart Retailing (Roy et al., 2017: 124–258)

Attributes	e-Retailing	Smart retailing
Space	Digital	Bridging digital and physical
Core technology	Websites	Innumerable sensors, smartphones, and apps
Nature of interactivity	Between customers and webstores: customer to customer	Customer to retailer Customer to customer Customer to products (brands) Products (brands) to retailers Machine to machine (touchpoint to touchpoint)
Nature of experience	Online shopping experience	New personalized and seamless customer experience emerges as a result of the nature of interactivity.
Service provision	Always-on services	Always-responsive services which is context specific

interactive and connected retail system which supports the seamless man-
agement of different customer touchpoints to personalize the customer
experience across different touchpoints and optimize performance over
these touchpoints". So, together with innovative technologies, interaction
and having a positive experience are main factors in smart retailing.

Improvements in industry 4.0 which include IoT, cyber systems and
cloud computing also stimulate smart retailing and increase productivity.
Fig. 3.3 shows the key features identifying smart technology for retailing
(Pantano and Timmermans, 2014: 104).

The concept of smart retailing goes beyond the application of a
modern technology to the retailing process by including a further level
of "smartness" related to the employment of the technology. Hence, a
smart technology for retailing, which generates the new concept of smart
retailing, can be investigated according to organizational and practical
dimensions (including the selling activity). Following these dimensions, six
main features are identified as the need for developing ad-hoc capabilities, changes in knowledge management processes (from and to clients),
the creation of smart partnerships (between retailers, clients, salesforce),

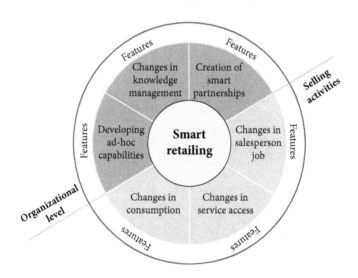

Fig. 3.3: Factors Identifying a Smart Technology for Retailing (Pantano and
Timmermans, 2014)

changes in service access, changes in salespersons' jobs, and changes in consumption (Pantano and Timmermans, 2014: 106–107). In Fig. 3.4, the key elements that contribute to creating the smart retailing framework are shown (Pantano et al., 2018). They are organization, economy, people, technology and market structure. *Organization* includes the people who are going to decide for a technology launch. It includes their attitudes to innovations or resistance to change, the fit of technology benefits to organization objectives, the internal and external resources for innovation and organization culture. *Economy* is an important driving force of smart retailing. It covers financial sustainability and economical environment. By supporting the company's growth and new business opportunities with smart use of the resources, smart retailing approach also contributes to sustainable development. *People* factor indicates the consumers in the smart retailing environment, retailer's management and employees who are in a smart cooperation in the retailing space. The fourth dimension is *Technology*, where innovative technologies are integrated in smart retailing. Smart retailing is highly related with the improvements in technology. These technologies are sophisticated systems such as real-time content softwares, hardwares and network technologies. They help the

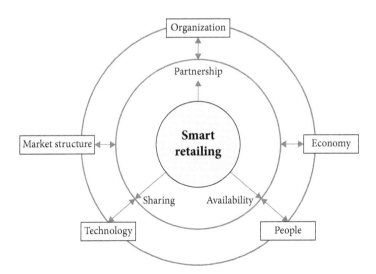

Fig. 3.4: Smart Retailing Framework (Pantano et al., 2018: 271)

retailers to collect customer data, predict the market trends, reduce the operating costs and effect the buying behavior of the consumer. Lastly, *Market structure* is determined by the number of competitors in the same sector and the innovation speed of the competitors. The market structure is related with the diffusion of smart technologies and number of competitors.

In the future shopping scenario of smart retailing shown in Fig. 3.5, during the shift from traditional to smart retailing access, connectivity, information sharing and collaboration increases. Going from traditional to smart retailing, more customer information is shared and collaboration between retailer and consumer increases. There is also an increase in the adoption of new shopping channels. For example, physical shop is used for "touching" and on-line channel is for "buying" (Pantano et al., 2018: 274).

In practice, smart retailing mostly involves artificial intelligence. Decentralization, optimization and transparency also play important roles in smart retailing. The outputs of smart retailing are efficiency, quality and sustainability.

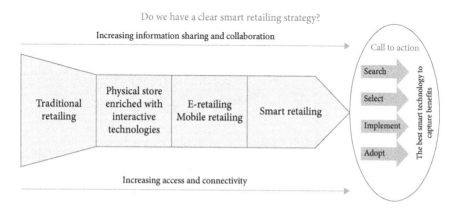

Fig. 3.5: Future Shopping Scenario Within the Smart Retailing Perspective (Pantano et al., 2018: 274)

3.3.1 Smart Technologies Used in Retailing

Analytics, artificial intelligence, augmented reality, virtual reality, sensors, facial recognition and cloud services are among the smart technologies that can be used by retailers to create competitive advantage. *Analytics* is key to understand consumers and provide personalized customer experience by analyzing demographics, customer traffic and other behavioral data. *Artificial intelligence* (AI) is automating and customizing the retail experience, aiming to make shopping easier for the consumer. For example, Amazon Go fully automated its market in Seattle and eliminated the checkout lines and cashiers. AR is an important element changing the shopping experience. Consumers may try clothes or test products virtually by AR. Sephora mobile application enables the consumers to test real-time virtual makeup. Shankar et al. (2011: 34) state that at Massachusetts Institute of Technology, an AR system was developed which enables shoppers to read packages and make informed choices at the stores using only hand gestures. According to Forbes, AR is changing the consumer retail experience; however, Virtual Reality (VR) is changing things in the retailer side. VR can be used for visualizing and redesigning stores and trying different layouts without having to physically rebuild the store. *Sensor data* is bringing great innovation to brick and mortar retailers. Leading brands put small beacon sensors in their stores and if the Bluetooth is on, customers are connected and the application of the retailer is downloaded. At the end, the retailer can track how long the customer stayed in the store and what she purchased. So, personalized discounts could be given to customers. Facial recognition technology may be used to understand customer preferences better. According to Forbes, facial recognition gives the information of which direction the customers are going and define the demographics. It can also prevent theft and help the retailer to design its store layout according to customer expectations (www.zdnet.com). Theft seems to be a retailer issue; however, it is also a customer issue. Its cost to retailers was $50 billion USD in 2016 and important portion of this cost was reflected to customers as higher prices (www.forbes.com). Lastly, cloud services are used for inventory tracking, stock availability, shipping details and orders. With the help of cloud computing, retailers can cut software development costs and process data faster.

Shankar et al. (2011: 35) classify promising new methodologies as biometrics (e.g. heart rate monitoring, and ambulatory EEG), RFID-based path tracking, eye cams, handheld scanners and infrared cameras. Omnichannel retail is another growing method which makes the customer experience superior by integrating on-line and offline channels. Digital coupons, virtual storytelling, e-mails and increased ads are also the new digital tools used commonly by marketers. Voice control has also become popular in retailing. Siri answers the consumer when she gives instruction to Google Home or with 'Theatro,' employees can communicate throughout a store via voice-controlled wearables (www.zdnet.com).

Since smart technologies help the retailers collect more and deeper data about consumers, it is a good opportunity for the retailers to build interactive and dynamic platforms for shopping. For example in services, customers and retailers may collaborate to use smart technology to bring "smart-partnerships" with the customers to achieve a high service quality (Pantano and Timmermans, 2014: 103). Interactive displays, smart shopping cards, RFIDs, shopping assistants, augmented-reality technologies in stores support the retailers by offering improved customer experience, better firm management, cost reductions, and increased business profitability (Roy et al., 2017: 257). By using these smart technologies, retailers and consumers will have the chance to reinvent and reinforce their role in the new service economy, by improving the quality of their shopping experiences (Priporas, et al., 2017: 375).

Roy et al.'s (2017: 267) research shows that smart technologies improve customer experience which consists of relative advantage, enjoyment, control, personalization, and interactivity. Positive customer experience creates customer satisfaction and decrease perceived risk related with the smart retail technologies. They also create competitive advantage for the retailer. Customer satisfaction also increases behavioral intentions, word-of-mouth intentions, stickiness to retailer, shopping effectiveness, and customer well-being.

3.3.2 How Do Brands Use Smart Technologies in Retailing?

In order to increase service quality, a new trend is to merge the retail channels. For instance, **M&S** e-boutique provides touch-screen displays

Fig. 3.6: Nike Id

that allow consumers to find and select their favorite products, access more information on items and pay by credit card. So, the consumers can reach more product range than the actual physical store stocks (Pantano et al., 2018: 268). **Walmart** lets the customers use their iOS devices to scan products to learn prices and customer reviews. In order to launch its new hiking boots, **Timberland** used a crowd-scale AR game in Madrid. The boot was dropped into the crowd and people tried to jump and throw the AR boot back to the air while the boot was reacting in real time to the movements. Since 2007, NikeiD service of **Nike** uses AR experience to support customers personalize their shoe (www.cmo.com). The new Nike iD Direct Studio (seen in Fig. 3.6) uses the technology to totally personalize a pair of Nike Air Force 1s in real time (www.gq-magazine.co.uk).

The smart technologies enabled retailers to personalize the physical and digital touch points. **Bonobon,** an online retailer, has opened a Guideshop (seen in Fig. 3.7) where the customers can be personally assisted to get the right products before online orders (www.cmo.com). Bonobos is a Guide that will help the customers to find perfect fit through the entire

Fig. 3.7: Bonobos

assortment and enable the consumer to walk out hands-free by shipping directly to home or office (www.bonobos.com/guideshop).

Especially transaction has many opportunities for innovative smart technologies in retailing. Point of sale and delivery are always ordinary parts of a customer journey. Self-order and self-pay kiosks in Fig. 3.8 from **McDonalds** seem to be more popular especially when combined with mobile wallets. This smart technology can make transactions faster and offer upsells (www.cmo.com).

With the popularity of digital touchscreens, food and beverage retailers will be able to offer digital ordering experiences as seen in Dunkin' Donuts example. This example in Fig. 3.9 is from **Dunkin' Donuts** (www.cmo .com).

Fig. 3.8: McDonalds

Fig. 3.9: Dunkin' Donuts

Fig. 3.10: Starship self-delivery Robot

Robotic delivery (from Starship in Fig. 3.10), which is a new tool today, seems to be a trend to rise in the close future. Starship is a company building a network of robots for self-delivery services (www.starship).

3.4 Future of Smart Retailing

It is vital for brands to understand that no technology can turn a negative brand-customer experience suddenly into positive. So, it is recommended that brands shall use these technologies to enhance customer relationships, get more traffic, loyalty and revenue (www.cmo.com). For 2019, the recommendations for smart retailing are (i) understanding the touchpoints of the customers if they are physical, digital or maybe both, (ii) delighting and surprising the customers so that they would like to interact with the brand more, (iii) making the journey of each customer unique (prepare a personalization strategy), and (iv) solve the problems especially in payment and delivery (www.cmo.com).

The new Internet of Things (IoT) technologies, such as indoor positioning, augmented reality, facial recognition, and interactive display enhance smart store implementations. Using these technologies, it is possible to cross machine data on customer behavior and direct interaction between sales staff and customers. With the data collected, stronger customer insights can be collected to improve shopping experience (Pantano et al., 2018: 94–95). IoT helps to increase the customer loyalty and sales, provides a personalized experience and enhances inventory management. Some smart technologies that offer opportunities for future retailers are smart shelves, beacons, robots, digital signage and face recognition. Smart shelves are innovative tools with an RFID tag, an RFID reader, and an antenna. The RFID tags are placed on the products and they transmit data to the RFID reader. This data is sent to IoT platform to be stored and analyzed. They are used in inventory management and CRM to avoid oversupply, shortage of goods, and thefts in stores. Also, smart shelves are useful tools to understand customer needs and preferences. Beacons are used to send automatic push notifications directly to customers. They use Bluetooth connections when the customers' smart device is in the coverage area. Beacons are mostly used for discount messages, promotions, in-store customer navigation, push messages and data collection. In order to design an engaging customer experience, beacons are good tools to use since they are small and easily attached to walls and counters. Many companies started to use robots in retailing and integrating robots in their work. For example, Target used robots in San Francisco to stock shelves and take inventory. Some restaurants in China replaced waiters with robots. With digital signage solutions such as IPad screens, digital-out-of-home applications (DOOH), intuitive touch screens, and in-store digital screens, retailers can offer personalized and interactive shopper experience. When compared with traditional media, the cost of this tool is less. By using the IoT, retailers can also push ads and price changes to stores in real time. IoT technology enables personalized promotions derived from the collected data. IoT will support the retailers in creating successful marketing campaigns, deliver superior service, improve inventory management and decrease costs (www.clickz.com).

Face recognition systems are included in personal devices such as iPhone X and this will make people adopt this technology and shopping styles of consumers may change. In the future, more than 1 billion devices will have face recognition in them. The vital question is "are the consumers ready for face recognition technology to be used for commercial reasons?"

Face recognition will bring important changes to retailing. In-store videos and face recognition systems are breakthrough technologies for retailers to collect demographic and behavioral data.

These systems also show how the consumer interacts with the in-store elements. The huge amount of data enables the retailer to analyze consumer's shopping behavior. It is easy to measure since the measuring platform and sensors are merged with the retail environment. The personal data of the consumer is not required since the measurement is anonymous. The analytical models help the retailer to convert consumer insights into successful marketing actions (Sharma, 2012).

While enhancing the customer experience, improvements in technology and innovations also create a new challenge. This is privacy. In both online and offline channels, huge customer data is collected. Especially the security of identity and financial information is critical for customers (Shankar

Fig. 3.11: Shopper Demographic Recognition Using In-store Video (Sharma, 2012)

vd, 2011: 39). It is a serious problem to sustain and protect the existence of this data from various effects. The solution to this problem is called as "cyber security" (Ünal and Ergen, 2018).

Cyber security is defined as "the protection of globally connected electronic data or equipment against criminal, unauthorized or accidental use and the technology and processes required to achieve this protection" (Coventry et al., 2014: 4). Cyber security is one of the major concerns that the customers perceive when smart technologies are the case. Many cyber attackers directly target human as the weakest part of the chain since the security softwares and operating systems are much stronger today. Bekoglu and Ergen (2016) state that consumers' attitudes towards these technologies and their concerns about privacy of personal information should also be considered by retailers while using these smart technologies.

Fig. 3.12: Shopper Gender Recognition Using In-store video (Sharma, 2012)

References

Bekoglu, F. B. and Ergen, A. (2016). "Reflections of Changing Consumer Trends and Technological Developments on Retailing". *International Journal of Research in Business and Social Science* (2147–4478), 5(2), 59–70.

Coventry, L., Briggs, P., Blythe, J. and Tran, M. (2014). Using behavioural insights to improve the public's use of cyber security best practices. gov. uk report.

Irfan, W., Siddiqui, D. A., & Ahmed, W. (2019). "Creating and Retaining Customers: Perspective from Pakistani Small and Medium Retail Stores". *International Journal of Retail & Distribution Management*, 47(4), 350–367.

Pantano, E. and Timmermans, H. (2014). "What Is Smart for Retailing?". *Procedia Environmental Sciences*, 22, 101–107.

Pantano, E., Passavanti, R., Priporas, C. V. and Verteramo, S. (2018). "To What Extent Luxury Retailing Can Be Smart?". *Journal of Retailing and Consumer Services*, 43, 94–100.

Pantano, E., Priporas, C. V. and Dennis, C. (2018). "A New Approach to Retailing for Successful Competition in the New Smart Scenario". *International Journal of Retail & Distribution Management*, 46(3), 264–282.

Priporas, C. V., Stylos, N. and Fotiadis, A. K. (2017). "Generation Z consumers' Expectations of Interactions in Smart Retailing: A Future Agenda". *Computers in Human Behavior*, 77, 374–381.

Roy, S. K., Balaji, M. S., Sadeque, S., Nguyen, B. and Melewar, T. C. (2017). "Constituents and Consequences of Smart Customer Experience in Retailing". *Technological Forecasting and Social Change*, 124, 257–270.

Savastano, M., Bellini, F., D'Ascenzo, F. and De Marco, M. (2019). "Technology Adoption for the Integration of Online–Offline Purchasing: Omnichannel Strategies in the Retail Environment". *International Journal of Retail & Distribution Management*, 47(5), 474–492.

Shankar, V., Inman, J. J., Mantrala, M., Kelley, E. and Rizley, R. (2011). "Innovations in Shopper Marketing: Current Insights and Future Research Issues". *Journal of Retailing*, 87, 29–42.

Sharma, Rajeev. (2012). "In-store Measurements for Optimizing Shopper Marketing", Chapter 17 from Ståhlberg, M., & Maila, V.. *Shopper marketing: How to Increase Purchase Decisions at the Point of Sale.* London and Philadelphia: Kogan Page Publishers.

Turano, M. AnnaMaria. (2012). "Tailing Your Shoppers: Retailing for the Future", Chapter 13 from Ståhlberg, M., and Maila, V.. *Shopper Marketing: How to Increase Purchase Decisions at the Point of Sale.* London and Philadelphia: Kogan Page Publishers.

Ünal, A. N. and Ergen, A. (2018). "Siber uzayda yeterince güvenli davranıyor muyuz? İstanbul ilinde yürütülen nicel bir araştırma". *Celal Bayar University Journal of Social Sciences/Celal Bayar Üniversitesi Sosyal Bilimler Dergisi*, 16(2), 191–216.

Vazquez, D., Dennis, C. and Zhang, Y. (2017). "Understanding the Effect of Smart Retail Brand–Consumer Communications via Mobile Instant Messaging (MIM)–An Empirical Study in the Chinese Context". *Computers in Human Behavior*, 77, 425–436.

Vrontis, D., Thrassou, A. and Amirkhanpour, M. (2017). "B2C smart Retailing: A Consumer-focused Value-based Analysis of Interactions and Synergies". *Technological Forecasting and Social Change*, 124, 271–282.

www.bonobos.com/guideshop, Accesed: 05.08.2019

www.clickz.com/the-future-of-retail-how-iot-is-transforming-the-retail-industry/214543, Accessed: 05.08.2019.

www.cmo.com/features/articles/2017/10/27/10-technologies-helping-to-overhaul-the-retail-experience.html, Accessed: 11.04.2019.

www.cmo.com/features/articles/2018/11/8/heres-how-retailers-are-adapting-to-digital-disruption.html#gs.4krxlq, Accessed: 04.08.2019.

The Deloitte Times (2018), Yeni nesil Perakendecilik, www2.deloitte.com/content/dam/Deloitte/tr/Documents/the-deloitte-times/haziran-2018_yeni_nesil_perakendecilik.pdf, Accessed: 04.08.2019.

www.forbes.com/sites/pamdanziger/2019/01/13/6-global-consumer-trends-and-brands-that-are-out-in-front-of-them-in-2019/#4a09e4c74fe4, Accessed: 05.08.2019.

www.forbes.com/sites/bryanpearson/2018/03/15/3-ways-retailers-can-use-facial-recognition-to-express-better-experiences/#235ebc017661, Accessed: 21.03.2019.

www.forbes.com/sites/bryanpearson/2018/03/15/3-ways-retailers-can-use-facial-recognition-to-express-better-experiences/#235ebc017661, Accessed: 21.03.2019.

www.gq-magazine.co.uk/article/nike-id-london, Accessed:05.08.2019.

www.starship, Accessed: 05.08.2019.

www.zdnet.com/article/10-technologies-leading-digital-transformation-in-retail, Accessed: 21.03.2019.

Erol Eryaşa

4. Cyberspace Impacts on Maritime Sector

Abstract: Maritime sector, which is an important part of global economy, consists of organizations managing infrastructures, ports, companies and ships. Maritime industry is taking its place with all its components in the digital transformation of Industry 4.0 and being highly dependent on digital and communication technologies. So, in maritime industry, which has high cyber security risks, being not able to manage cyber risks and increase the cyber security awareness at all levels may cause serious loss. The aim of this section is to describe the effects of digitalization on the maritime sector, to identify cyber risks and vulnerabilities related to maritime, to discuss the reactions and sectoral regulations of various maritime organizations.

Keywords: Maritime, Cyber Attacks, Vulnerabilities

4.1 Introduction

In the developing world, the first five of the fastest-growing sectors in 2017 are renewable energy, virtual reality, biotechnology, cyber security and artificial intelligence. Among them, cyber security is in high demand and experts estimate that the global market will be worth $165.2 billion by 2023 (World Finance). Cyber Security and the maritime sector have confronted each other together with developing technology to transform the processes in the sector in order to increase the efficiency of maritime transportation.

The maritime sector has always been one of the best practice areas of current technology as the most important sector that provides for the expansion and integration of countries. Maritime transportation has been the most preferred solution in international trade by providing more than 80 % of the world trade. About 50,000 ships and 1 million seafarers are actively involved in this worldwide trade (ICS). In other words, the maritime sector is a vital part of the global economy and constitutes the basis for the effective functioning of all aspects of modern society.

According to the result of the World Economic Forum's Global Risks Perception Survey (GRPS), in terms of likelihood of first ten risks, the first three risks are related with the environmental problems. Data fraud/

theft and cyber attacks are ranked respectively as 4th and 5th (World Economic Forum, 2019: 3). Cyber security risk has become more widespread and its destructive potential has increased in recent years with the increasing number of cyber incidents that were once considered extraordinary. GRPS underlines that in 2019, the risks of cyber attacks and data fraud followed by economic, commercial and political concerns are going to be among the top five global risks in the world (World Economic Forum, 2019: 12).

Until 2010, despite the current level that cyber security reached, the majority of cyber attacks were carried out as an attempt to obtain personal or financially sensitive data. Today, the nature of the threat has changed, and through the acquisition of industrial control systems, all sectors have begun to experience very complex attacks, which try to harm material and reputation. In 2021, the total cost of cyber security crimes to the world economy is expected to rise to $6 trillion annually (Cybersecurity Ventures).

The maritime sector, covering a wide range of international organizations, ports, companies, ships and all stakeholders, has entered a new phase of its evolution in parallel with developments in information technology. In addition to the efficiency and facilities achieved by the maritime industry's rapid progress towards a digital world, it adds serious complexity to the operational environment of ships, ports and ships. Considering both the intense use of technology in maritime transportation and the ratio of maritime transport in world trade, it can be seen that a possible failure may affect the entire supply chain and cause costs increases and loss of reputation. In this context, ports and ships are critical infrastructures with information technologies integrated into industrial control systems. For this reason, security and cyber security awareness studies in maritime are

"Maritime cyber-attacks are no longer the stuff of science fiction. They are happening now and the threats are growing".

Fred Roberts,
Professor of Mathematics and Director of Command, Control and Interoperability Center for Advanced Data Analysis

becoming more and more important. In the maritime sector, where cyber security risks are high, it is absolute that cyber risks cannot be managed well and that if the level of cyber security awareness of employees at all levels cannot be increased, serious losses may occur. In the following parts, cyber definitions, effects of digitalization on the sector, management of cyber risks, categorization of cyber threats, cyber attacks and sectoral regulations will be discussed.

4.2 Cyber Definitions and Their Counterparts in the Maritime Sector

In order to comprehend the impacts of the cyberspace in the maritime sector, it is necessary to mention cyber definitions and their counterparts in the sector.

Cyber risk means the risk of damaging the financial performance and reputations that may occur due to the loss of commercial information, accidents, incidents, and business interruption in the information technology system (IRM, 2014: 10). **Cyber attack** in the maritime sector is "*an attempt to damage, disrupt, or gain unauthorized access to a computer, computer system, or electronic communications network (IMO 1:4), to ships, terminals, ports and all computerized equipment supporting marine operations*" (Connecting EU). *Cyber security covers either the whole of the measures to be taken against cyber attacks or the elements including tools such as connected devices, personnel, infrastructures, applications, services, transmitted, processed or stored data and information in the cyber environment (ITU, 2008: 8). Cyber security are the combination of tools, policies, security concepts, security measures, guidelines, risk management approaches, actions, training, best practices, assurance and technologies that can be used to protect the cyber environment and user assets (IET, 2017: 14).* This **cyber environment** in the definition can be evaluated as the intersection of the marine environment with cyberspace and as the interconnected networks of both IT and cyber-physical systems that use only electronic, computer-based and wireless systems including information, service, social and business functions in cyberspace. Shortly, **cyber security in the maritime field** can be understood as the protection of electronic systems, communication networks, control algorithms, softwares, users and the data in the maritime infrastructure against malicious attacks, damage, unauthorized access or manipulation (Garcia-Perez et al., 2017: 2).

Cyber resilience is another related term with maritime field. It actually means the ability to prepare, respond to and recover from cyber attacks. It is closely linked to the safety and availability of critical systems in ships. These systems are constantly monitored to provide situational awareness based on data from various sensor types. Therefore, the integrity, accuracy and availability of such data are critical to the safe and efficient operation of the ship systems, which are dependent and integrated. It is essential to maintain the integrity of all systems and thus to maintain cyber resilience to understand the relationships between these dependencies and systems. Besides, the ongoing process of information update regarding cyber security is necessary to increase cyber resilience in the maritime industry (IET, 2017: 8).

4.3 Effects of Digitalization on Maritime Sector and Cyber Security Requirement

Digitalization that means the use of digital technologies to change a business model and provide new revenue and value-generating opportunities continues to transform all industries and services, including the maritime sector. Maritime sector is taking its place with all its components in the digital transformation process under Industry 4.0 and is becoming increasingly dependent on the widespread use of digital and communication technologies. The necessity of developing and optimizing different business models in today's maritime is directly linked to the implementation of new digital technologies that enable automation of processes and functions on ships. In order to embrace digitalization, companies implement different technologies and try to combine them with new business models, services and products. At the end of this process, different operational situations occur for the whole sector (for companies, ports and ships).

The main reasons of increase in the popularity of digitalization in the maritime sector are costs and functionality; increased computing power and speed, increased data storage capability, improved connectivity capability and real-time data from sensors (Lagouvardou, 2018: 20–21).

Thus, as digitalization and systems become more interdependent and more connected, new areas are rapidly emerging in response to the latest demands. The technologies that are expected to have the greatest potential impact on the maritime sector over the next decade are seen as "big data", "blockchain technology" and "Internet of things". The maritime industry gives high importance to these three new concepts and seem to be relatively well prepared to face them (GMF, 2018: 14). Regarding **big data**, there are many benefits that maritime transport can achieve. An approximately 100–120 million data points are generated daily from different sources such as ports and ship movements. Companies can analyze these data to identify activities such as route or port preferences. Ultimately, a serious performance increase can be achieved by using big data (Trelleborg Marine Systems, 2018: 6). When **blockchain technology** is considered, the number of projects targeting the maritime industry is increasing rapidly with new initiatives in maritime insurance, fuel purchase and port management triggering all block chain-based processes to increase productivity and reduce costs (Safety4sea, 1). Many companies, such as Maersk and IBM, use blockchain technology to become competitive in global digital trading platform (Maersk). Lastly, **Internet of Things** will provide many benefits, from reducing fuel consumption to more efficient monitoring of cargos. By investing in sensors and systems, maritime companies are expected to reduce their costs of operation and maintenance, and on average, the IoT-induced cost saving rates are expected to be 14 % in the next five years (Safety4sea, 2). The Rotterdam Port Administration has launched the Internet of Things (IoT) platform to better plan and manage maritime transport. In this way, waiting time reduction, berthing, loading/unloading and take-off times can be optimized (Digitalship 2: 16). In addition to these new concepts, 3D printing, artificial intelligence, autonomous technology and robotics can also be listed as important technological milestones in maritime sector. For example, Google and Rolls-Royce are working on projects on autonomous maritime transport and intelligent systems (Rolls-Royce).

While the digital evolution of the maritime industry offers various opportunities to improve the way maritime companies do business, the cyber risks are a concern for many maritime managers. Cyber security became a vital factor that needs to be managed carefully.

> *"Digitalisation of shipping is 'underway', but needs
> to be steered at safe course and safe speed".*
>
> Capt. Erol Eryaşa

According to the results of the survey conducted by Reed Smith with participants from maritime industry of various markets and regions in 2018, in the next five "analytics of big data" will be the most important impetus of change. Also, the greatest challenge the maritime industry will be facing in the next five years will be "cybercrime" as third after "flattening demand growth" and "operational difficulties" (ReedSmith, 2019: 5–9). These findings show that the maritime industry has to move rapidly in order not to be exposed to cyber attacks, which may have serious economic and reputation-based consequences. Also, cyber security awareness must increase and efforts should be made to take all measures to protect against cyber attacks at all levels from global companies management levels to individuals. However, the research finding shows that the maritime industry is not ready to face with the digitalization and it seems that the implementation of international regulations is not fully effective at present. The maritime industry is vulnerable to cyber threats (Digitalship 1: 23).

4.4 Management of Cyber Risks and Vulnerabilities

In a rapidly evolving maritime environment, cyber security needs to be integrated with a holistic approach from international organizations to countries, companies, ships and individuals in order to respond to new and continuously self-renewing cyber risks and provide sustainable durability. To manage cyber security risks effectively, it is necessary to define these risks. Risk management, which is essential for safe maritime transport, is mainly focused on operations in the physical field, but with technological development, more dependency on digitization, integration, automation and network-based systems has created an increasing need for cyber risk management in the maritime industry. The increasing use of digitalization due to the benefits provided for the maritime sector, and in particular the information technology (IT) and operational technology (OT) elements on ships, are more connected to each other and to the Internet, thereby increasing the risk of unauthorized access or malicious attacks to ships' systems and networks. Since ships make decisions based on the integrity of the data which they receive through existing IT and OT systems, the cyber risks that could cause significant damage to the maritime industry should be carefully evaluated.

IMO (International Maritime Organization) defines **cyber risk management** as "*the process of identifying, analyzing, assessing and evaluating cyber risk to a reasonable level by taking into account the costs and benefits of actions taken by stakeholders*" (IMO 2). In this respect, Safety Management System (SMS) is published by the companies in order to provide a safe and efficient working environment by creating appropriate safe practices and procedures based on the evaluation of all risks identified by the personnel on board and the environment. According to this plan, the company needs to assess the risks arising from the use of IT and OT systems on board and provides appropriate protection against cyber incidents.

SMS includes instructions and procedures to ensure that ships operate safely and the environment is protected in accordance with relevant international and flag state legislation. When preparing these instructions and procedures, the risks arising from the use of IT and OT on the ship should also be considered taking into account the relevant codes, guidelines and recommended standards. Briefly, OT systems control the physical world and IT systems manage data (BIMCO, 2018: 5). OT is hardware and software that directly monitors/controls physical devices and processes. IT covers the range of technologies used in information processing, including software, hardware and communication technologies. OT systems are different from traditional IT systems, but via the Internet, OT and IT systems approach each other. The disruption of the operation of the OT systems can pose a significant risk to the safety of the personnel on board, to the cargo, to the marine environment and to prevent the operation of the ship. Cyber risks and security issues will increase in the maritime sector with the displacement of cyber risks from information technology (IT) to operational technology (OT) (DNV GL 2). Moreover, when the shore-based and ship-based Information Technology (IT) systems are interconnected, the attack can make the vessels very vulnerable to cyber attacks. It is possible to attack the company's shore-based IT systems without the need for direct attack of ships, and that the vessel can be easily entered into critical Operational Technology (OT) systems.

Actually, a loss of IT systems on board is a business continuity issue and should not have any impact on the safe operation of the ship. However, the loss of OT systems may have a significant and immediate impact on the safe navigation and operation of the ship (BIMCO, 2018: 34).The segregation of IT and OT systems and their risks stated in Tab. 4.1.

Effective cyber risk management should provide an appropriate awareness of cyber risk at all levels of an organization. Awareness and preparedness level should be in line with the roles and responsibilities of the cyber risk management system. Functional concepts that support effective cyber risk management are (i) definition, (ii) prevention, (iii) detection, (iv) responding and (v) recovery. These functions should be handled

Tab. 4.1: IT & OT Systems and Their Risks (Rossi, 2018: 19)

Titles		At risk
IT Information Technology	IT networks	Mainly finance and reputation
	E-mail	
	Administration, accounts, crew list vs	
	Planned Maintenance	
	Spares management and requisitioning	
	Electronic manuals	
	Electronic certificates	
	Permit to work	
	Marine trading letters	
OT (Operation Technology)	On-board measurement and control	Life, property and environment + all of the above
	ECDIS	
	Power management	
	GPS, CCTV	
	Remote support for engines	
	Data loggers	
	Engine control	
	Dynamic positioning	
	PLCs	

interactively and continuously (IMO, 2017: 3). *Definition* is the identification of roles and responsibilities of both shore personnel, crew members (captains and officers) and key staff for cyber risk management and determining the systems, assets, data and facilities that constitute a risk for ship operations when interrupted. This concept also means defining all known weak points that are likely to give a deficit against an attack (UNCTAD, 2018: 85). *Prevention* is implementation of risk control processes, measures and emergency planning to ensure continuity of maritime transport. *Detection* is development and testing of procedures to detect a cyber event on time. Detection and reduction of cyber security attacks can only be possible by training and awareness of all staff. *Responding* is implementation of a well-prepared response plan for solving problems for personnel, systems and normal operations. In this context, it is the development and implementation of procedures to ensure cyber resilience and to restore corrupted processes or services due to a cyber incident. Response and recovery plans should be reviewed and implemented at regular intervals (Secure State Cyber). *Recovery* is implementing measures to back up and restore the cyber systems required for maritime transport operations affected by a cyber incident (IMO, 2017: 3).

While defining systems that pose a risk for ship operations, it is necessary to mention the operation and management of numerous systems of critical importance for the safety of maritime transport and the protection of the marine environment. Vulnerabilities resulting from accessing, interconnecting, or networking to these systems appear to be important cyber risks (IMO, 2017: 1).

The increasing digital complexity of the ships and the connection of the Internet to the services provided by the coastal networks cause more cyber attacks. The systems on board may be the target of a direct cyber attack or may be vulnerable as a system affected by a successful cyber attack. In

---

"We are vulnerable in the military and in our governments, but I think we're most vulnerable to cyber attacks commercially. This challenge is going to significantly increase. It's not going to go away".

Michael Mullen—US Navy Admiral Chairman of the Joint Chiefs of Staff

general, stand-alone systems are less vulnerable to cyber attacks, uncontrolled networks, or those directly connected to the Internet. Therefore, care must be taken to connect ship systems to uncontrolled networks. In doing so, it is important to note that there are many human factors at the beginning of many cyber events (BIMCO, 2018: 13).

There are three network systems onboard to support safe navigation and operation, and to meet administrative needs: a "navigation network" that supports the safe navigation of the ship; an "engineering network" that controls the ship's propulsion system, along with the machine system and auxiliary systems; and the "administrative network" supporting the operations and personnel needs. Cyber-vulnerable systems that work in these networks can be listed as follows (IMO, 2017: 1):

- Bridge systems;
- Cargo handling and management systems;
- Propulsion and machinery management and power control systems;
- Access control systems;
- Passenger servicing and management systems;
- Passenger facing public networks;
- Administrative and crew welfare systems; and
- Communication systems.

Ports and maritime land-based infrastructure facilities can also be the target of cyber attacks because of their Internet connection. The list as cyber vulnerable systems for ports are vulnerable IT structures included in ships, cargo-handling systems (ICS, SCADA), cargo management systems (RFID, Tracking systems, inspection systems), passenger management systems, corporate IT systems (e-mails, intranet, etc.), ground systems for vessels (mooring, tracking, maintenance), alarm and access control systems, personnel and stevedore management systems (Sofronis, 2018: 5).

Nowadays, there are many devices with advanced technology used as bridge navigation devices. These are ECDIS (Electronic Chart Display and Information System), AIS (Automatic Identification System), GNSS (Global Navigation Satellite System GPS), Radar/ARPA (Radio Detection and Ranging/Automatic Radar Plotting Aid), Steering System (Computerized Automatic Steering System), VDR (Voyage Data Recorder–"Black Box") and GMDSS (Global Maritime Distress and Safety System) (Zăgan et al.,

2017: 222). These are great convenience for watchkeeping officers. Increased use of digital, networked navigation systems, coastal systems and networked interfaces for updating and delivering services make such systems vulnerable to cyber attacks. Removable media are equally at risk from systems that are not directly connected to the network. The cyber event that may occur may result in denial of service or manipulation, and may affect all bridge systems. Therefore, it is possible for these devices to be exposed to cyber attacks by intentional, unintentional or natural reasons, and to be able to mislead the captain and officers by off-ship intervention and to have significant consequences that will lead to drift off ship's route.

Several researchers have reported that the technology used for navigation at sea, such as GPS (Global Positioning System), AIS (Automatic Identification System) and ECDIS (Electronic Chart Display and Information System), has significant weaknesses in terms of cyber security, and that each is vulnerable to potential attacks (Zăgan et al., 2017: 222; Saygılı and Ünal, 2018: 248). In fact, the ECDIS system is not a sensor in itself, but it is integrated with many sensors and systems. Therefore, the vulnerability is high. The merchant ships become a soft target for cyber attacks because of the weak signals of GNSS (GPS) system that lack encryption or authentication (Fahey, 2017). GPS and ECDIS are usually integrated with AIS, which is considered a vulnerable target for cyber attacks. Thus, the proven hacking and ease accessing of these systems give cybercriminals the opportunity to create impacts in world trade by disabling ships while underway. There are four main practices to acquaint with ECDIS to malware, three of which relate to the introduction or update of electronic charts (ENCs). The first method is that the malware is infected with a USB memory device inserted in ECDIS. The second one is using an infected CD for ENC updates. The third is using a direct connection between the ECDIS and the ship's satellite communication. The fourth method is an indirect connection between satellite communication and ECDIS via another operational technology (OT) device, such as the AIS (Automatic Identification System) or the VDR (Voyage Data Recorder) (Marinemec). AIS devices and systems are widely used in marine transport for VHF band operation and the ship's static, dynamic and voyage information are shared. Therefore, AIS systems are widely used in navigation, traffic monitoring, collision avoidance, search and rescue operations, accident investigation and piracy prevention,

maritime safety operations and are vulnerable to cyber attacks (UNCTAD, 2017: 86). It is possible for the pirates to see the type, position, direction and speed of the ship in range using this weakness and to develop an attack plan by seeing their cargoes. In any case, an incorrect "Collision" warning can be triggered in the Radar ARPA system by creating the risk of collision (CPA Spoofing) and a fake weather report can be generated to guide the ship to alter course (Faking Weather Forecasts); with AIS-SART (Automatic Identification System-Search And Rescue Transponder) deception an attacker may create false signs of danger, be directed to the marine areas controlled by them, or be prevented from taking on the identity of the authorities and disabling the ship's AIS device (AIS-SART Spoofing) (Balduzzi et al., 2014: 438).

Although the cyber-vulnerable systems in ships are predominantly bridged systems, they also include the following systems.

Cargo handling and management systems: Digital systems used in cargo management and control, including dangerous cargoes, can interface with various systems on land. Such interfaces make the cargo management systems and cargo declarations vulnerable to cyber attacks (BIMCO, 2018: 13).

Propulsion and machinery management and power control systems: The use of digital systems in the control and monitoring of machinery, propulsion and steering systems on board creates cyber vulnerability. The vulnerability of these systems can increase when used in conjunction with remote condition-based monitoring and/or are integrated with bridge equipment (BIMCO, 2018: p.13).

Access control systems: Surveillance is digital systems used for supporting access control to ensure the physical safety and security of a ship and its cargo, including ship security alarm and electronic "personnel-on-board" systems (BIMCO, 2018: 13–14).

Passenger servicing and management systems: Digital systems are containing the information of passengers (BIMCO, 2018: 14).

Passenger facing public networks: Wired or wireless is installing for use by passengers must be considered uncontrolled and should not be connected to critical safety systems on board (BIMCO, 2018: 14).

Administrative and crew welfare systems: Computer networks that are used by ship management or staff on board should be considered particularly vulnerable to uncontrolled attacks due to Internet access and e-mail. So, they should not be linked to critical security systems on board (BIMCO, 2018: 14).

Communication systems: The availability of Internet connection for satellite and/or other wireless communication may increase the vulnerability of the ships. Cyber measures applied by the service provider should not be trusted alone (BIMCO, 2018: 14). On the other side, the most important element in the vulnerabilities related to the above-mentioned networks, systems and devices is the human factor such as company personnel, port employees, ship crew, managers and passengers. In particular, crew can usually connect their hard disk or USB stick to a computer inside the ship system. Hackers may be infected intentionally or unintentionally in the hope of connecting the device to crew's specific equipment, the device in the company or ship network, and malware may then automatically spread to all or part of the IT system. Cyber risk increases if legacy hardware will no longer be compatible with current security software. Furthermore, the fact that the ship personnel is changing in short periods and that the personnel has different levels of awareness. This can be considered as weakness in terms of cyber vulnerabilities. Together with IT and OT systems mentioned above, human factor holds important places for cyber security vulnerabilities. Because human failures may make cyber attacks successful (Daum, 2019: 7).

4.5 Categorization and Analysis of Cyber Attacks

In order to take necessary security measures, it is necessary to have knowledge about the types of cyber attacks and how they are made. Cyber attacks target software, hardware and infrastructure running in cyberspace. The period of preparation of cyber attacks depends on the motivation and goal of the attacker, as well as the cyber awareness and resilience of the company/ship. Attacks generally follow the following steps (International Armour Co):

- **Reconnaissance:** It is the first stage of an attempted attack and is the process of gathering information from the company, ship, port, seafarer from open sources to be used in cyber attack.
- **Delivery:** At this stage, the attackers attempt to access the ship system and data of the company. These may be through online services, cargo

tracking systems, or malicious files to the seafarers, by sending e-mail with links, or by malware infection as part of the update of the software used on the ship.

- **Breach:** Depending on the cyber vulnerability of the system of the ship or company to be determined by the attacker, for instance, it may interfere with the ECDIS or GPS system, it may make changes that affect the operation of the system or access the crew list (or passenger list), cargo manifest or provide access to the machine control system. It is important that the breach made by the attacker does not cause a significant change in the state of the device.
- **Affect:** Depending on what the attacker would have had on the company or ship system and its data, he/she could access and manipulate commercial, crew, cargo and passenger information, or interfere with the normal operations of the company or ship.

Cyber attacks can also take place on the ship as targeted (Tab. 4.2) or by accessing personnel on board with conscious malicious action or unconsciously by neglect or ignorance.

- **Brute force:** An attack that attempts many passwords in the hope of guessing correctly. The attacker systematically checks all possible passwords until they are correct.

Tab. 4.2: Categorizing the Threats (Lagouvardou, 2018: 20–21)

	Intentional	Unintentional
	Brute force	
Targeted	Distributed Denial of service (DDoS)	Falling victim to social engineering
	Spear-phishing	Escaped proof-of concept
	Subverting the supply chain	Runaway pentest
	Port scanning	
	Malware	
Untargeted	Phishing	User Error
	Water holing	
	Port scanning-	

- **DDoS:** It is designed to prevent authorized users from accessing information and to deactivate the target system for a certain period of time with continuous traffic flow from different sources to the target network. A DoS attack will eventually lead to deliberate crashing of the targeted IT system and economic losses. The most important difference of DDoS attack from DoS is that the attack is carried out from multiple points to a single center. DDoS attacks on systems used in maritime transport will hinder access to these systems, which will make the management of ship traffic difficult and sometimes even unmanageable (Keleştemur et al., 2017: 11).

- **Spear-phishing:** It is a more advanced and dangerous version of phishing. It targets personal e-mails such as phishing, but includes malicious software or links that automatically download malware. However, in this method, not only e-mail also social media channels also allow the attacker to gather information about the target for the Spear-phishing (Pajunen, 2017: 21 as cited in Palo Alto Networks, 2016).

- **Subverting the supply chain:** It is a type of attack involving equipment, software or support services delivered to the company or ship. This type of attack is extremely popular in the maritime industry due to the real-time connection between the members of the supply chain, ports or terminals. It is a way of manipulating other companies (Lloyd's Register).

- **Port scanning:** The main purpose of this attack is controlling the ports that the user opens for incoming connections.

- **Malware:** It is the general name of many threats on the Internet and is the abbreviation of malicious software. There are various types of malware such as Trojans, spyware, ransom software, viruses and worms. It is designed to provide or damage a computer, server, or network without the victim's knowledge. In the past, it was easy to recognize it as an independent virus that only transmitted and reproduced machines, but nowadays, malware may be difficult to recognize because it can be updated to prevent detection by traditional malware blockers. Malware can also be tailored to specific individuals or organizations (Pajunen, 2017: 21 as cited in Palo Alto Networks, 2016). The main purpose of Malware is to steal resources from the computer using known deficiencies and problems of the computer and the network (e.g. an outdated software).

- **Social engineering:** The technique used by cyber attackers to allow insiders to violate security procedures through interaction with social media (Lloyd's Register). For example, the passwords that most people will not provide when requested are given by a seemingly reliable person, such as a help desk employee or network administrator (Pajunen, 2017: 21 as cited in Walker 2012).
- **Phishing:** Phishing attack is designed to send an e-mail containing links to a fake website or to download malicious content to a large number of potential targets and to spoof the user into exchange for confidential/private information. The e-mail the user gets looks like an official organization, however, when the user clicks the links, the attacker gets all the information the user enters into the counterfeit website (Pajunen, 2017: 21).
- **Water holing:** Water holing is to create a fake website, or use a real website to fool visitors (Lloyd's Register). In order to determine the type of websites they often use, the attacker first determines the company's employees as targets. The attacker then searches for vulnerabilities on websites and injects the malicious JavaScript or HTML code that directs the target to a separate site where malicious software is hosted. This compromised website is now ready to infect the target with malware injected when it is accessed (Pajunen, 2017: 22 as cited in TechTarget 2015).

A cyber security survey was conducted in 2018 with the cooperation of BIMCO (Baltic and International Maritime Council) and "IHS Fairplay" for the analysis of cyber threats. In this study, different types of threats were questioned by the maritime industry. Phishing (49 %) and malware (44 %) were the most common threats (IHS Markit, 2018).

4.6 Cyber Risks and Cyber Attacks in the Maritime Sector

In the maritime sector, cyber attacks occur in different targets and dimensions. The ships, maritime enterprises, port and infrastructure facilities can be targeted. The oldest known cyber-maritime incident occurred in 2001 when a teenager attacked the computer system at Houston port after the denial of service (Gliha, 2017: 229 as cited in B. van Niekerk). This attack was a clear warning to the maritime industry to raise awareness of cyber threats. However, due to insufficient measures, the number of attacks increased. So, companies tended not to report cyber accidents not to damage their reputation or being attacked (Gliha, 2017: 230 as cited in P. Glass). In other words, it is clear that number of real attacks are much more than the recorded attacks (Gliha, 2017: 230 as cited in CyberKeel).

Before 2010, cyber-attacks were usually in the form of hacking, sending spam e-mails, changing the content of the website, or denial of service. Nowadays, cyber attacks are no longer carried out as large-scale attacks that have been developed for a specific purpose, such as STUXNET. Instead, theft of information such as Petya, NotPedya, WannaCry, Equifax targeting companies and ships, blocking access to the computers by locking the hard disk, or disabling OT systems such as AIS, GPS, ECDIS, and so on, or damaging and misleading the ships and companies with false messages.

In 2010, the malware called the Stuxnet worm was used to infect and disinfect the systems of computers connected to the centrifuges at an uranium enrichment plant in Natanz, Iran. The attack occurred after about 1000 centrifuges were damaged in the plant (Lagouvardou, 2018: 41). Following the success of the Stuxnet worm, it was found that targeting a plant with malware rather than infecting malicious software into the energy system could lead to dangerous problems. According to a report by the security and antivirus company ESET, it has been found that this malware is then widely distributed to users in computers in Iran and the surrounding countries, and inadvertently spreading them, and Worm has manipulated computer systems to manipulate and destroy most of the centrifuges (*Houston Chronicle*).

Cyber attacks on ships, which are the main factor in the maritime sector, started to come to the agenda of the world public in 2011 with the maritime piracy activities. The target of the attacks are the identification and navigation of ships naturally. The AIS (Automatic Identification System), which has been used mandatory by the IMO countries since 2005, is open to access and thus cyber-vulnerable systems, which operate on the VHF band and share many information about the ship. Pirates know how to use AIS to detect ships arriving in range and even look up ship information to see their valuable goods or cargos. For this purpose, IMO has allowed the AIS to be closed if the operation of the AIS (Automatic Identification

System) is considered by the ship's captain to endanger the safety of the ship in the maritime areas where pirates and armed robbers operate (IMO, 2015, Resolution A.1106 [29]: 8). As a precaution, LRIT (Long Range Identification and Tracking) system, which has no distance limitation and works with satellite systems, has started to be used. The LRIT system is not available for open access and may be used by the authorized authorities and the contracting countries of the IMO Convention. In addition, in that region, the armed special units started to protect the ships and due to this reason the piracy activities against the ships were developed and carried out with cyber attacks. Somali pirates and their accomplices, especially foreign trade organizations, began to work with experts who knew how to get information that could help them target vulnerable and valuable ships, how ship owners and shipping companies would get into their 'secure' computers, and how to obtain non-public information including the insurance of ship and cargo. For example, at the end of 2011, the MT Enrico Ievoli which has Italian flag was carrying 15,750 mt of caustic soda and navigating from the Persian Gulf to the Mediterranean, and was targeted in a premeditated way. Her voyage plan, cargo and crew, location, and the fact that she didn't have armed guards were all known in advance by her Somali attackers, thanks to the help from the Italian mafia, which commissioned the hijacking. She was hijacked on the coast of Oman while he navigate her way to the appointment with the foreign navy convoy who was patrolling and assuring the security of the Gulf of Aden corridor (Safety4sea3).

In 2013, Researchers at Trend Micro company have demonstrated how AIS can be endangered by staging fake emergencies and exposing "ghost ships" from the system, causing other AIS users to see the ship in the wrong place. It has likewise shown that a ship can alter its nationality or cargo to ensure that ships carrying dangerous cargoes appear harmless (TrendMicro).

In the other way, by deception on the GPS device, it is possible to take the ships out of their routes and direct them to the area where the pirates are located. If "situational awareness" is not developed due to the automation of deck and machine components of a state-of-the-art vessel, it may not be immediately intervened by ship personnel. In 2013, cyber security experts proved how easy it was to enter a ship's navigational equipment when the route of a super yacht was intercepted and the ship's course changed. The GPS is a device required to determine the position of the ship, as seafarers are well aware, and ensures that it works correctly by providing position information for the devices in the ship navigation system. In 2013, Todd Humphreys and his students from the University of Texas have conducted a test to deceive the ship's GPS system. By replacing the original GPS signals with fake signals sent by a $ 2,000 spoofer (GPS signal mixer), the White Rose of Drax made the navigation equipment of the yacht think that the ship was out of three degrees (Pajunen, 2017: 13 as cited in Psiaki and Humphreys 2016). Worst of all is that the GPS device cannot tell if it is simulating, if the GPS signal has not been blocked (jamming) or distorted, the forgery will not cause an alarm on the navigation device.

> *"Jamming just causes the receiver to die, spoofing causes the receiver to lie".*
>
> David Last, Consultant,
> Former President of the UK's Royal Institute of Navigation

Since the original GPS signals come from various satellites in different directions, a deception signal is likely to come from a single source. From this fact, GPS manufacturers have taken the necessary measures with the software update. This experience shows that although ships have navigation systems and sensors in accordance with technology, ship navigation largely depends on satellite systems, and a deception attack, such as the effects of a slowly changing ocean stream, can be hidden (Sea Knight Maritime Inc).

In April 2015, South Korea exposed one of the most significant GPS jamming incidents when over 1,000 aircraft and 250 ships were affected. Because of the attack on GPS, the ships were in the unknown position and forced back to the port. The incident has been claimed to come from state-sponsored actors in North Korea (Lagouvardou, 2018: 86). Such attacks benefit from the fact that civilian GNSS (unlike military GPS) waveforms are unencrypted and unauthenticated.

In June 2017, approximately 25–30 ships, which were located in the Black Sea off Novorossiysk harbor, appear at the Gelendzig Airport about 32 kilometers inside at the same time. Such incident has shown that it can be intervened remotely from the navigation systems (NewsScientist).

The target of cyber attacks was not only ships and other stakeholders involved in the maritime sector. The terminals in several ports, including the Rotterdam Port of the Netherlands, Jawaharlal Nehru Port which is the largest container port in India and including terminals in the United States were attacked and delayed in shipments due to closure of the terminals in the port (UNCTAD, 2017: 88).

Zombie Zero, a method of attack discovered by TrapX in 2014 to attack transportation and logistics companies, was used by attackers in the form of using malicious software covertly placed on new barcode readers. In China, at least eight maritime transport and logistics companies have been damaged due to a malware that has already been installed on newly produced scanners (Gliha, 2017: 232).

The change of cargo manifestos made it possible for illegal loads such as drugs and guns carried within the container to be shown as an ordinary and non-dangerous burden. For example, in Port of Antwerp, the port systems are hacked and the load information is deleted from the system in order to steal the load and to delete the track. Starting in June 2011, drug traffickers with hackers have hijacked IT systems that control the movement and location of containers in the port of Antwerp, Belgium. For two years, hackers have organized cyber attacks that control the movement and location of containers in the Port. These containers were used by drug traffickers to store and transport illicit drugs (Seatrade). Although the event was experienced in 2011 due to the success of the methods used by the attackers, it took one year to be noticed and the event was reflected on the media in 2013. It has been revealed that attackers have been infiltrated into the system for about a year (Keleştemur et al., 2017: 10).

In 2015, the MODU (mobile offshore drilling unit) staff unintentionally downloaded malicious software that spoiled 'computer networks' by directly downloading infected files of pornography and illegal music and bringing infected laptops and USB drives. Malware has rendered dynamic positioning thrusters non-functional and so MODU was dragged from the well zone. That case revealed the weaknesses of MODUs in terms of cyber security (LinkedIn).

In November 2017, Clarksons, one of the largest shipping agencies in the world, was attacked. A judicial investigation revealed that an unauthorized third party had access to some Clarksons computer systems in UK between 31 May 2017 and 4 November 2017, copying the data and requesting a ransom for a safe return. This event indicates that the cyber security issue in the maritime industry shows that not only floating or mobile elements will affect all stakeholders, including ports, companies, insurers, brokers, suppliers, class organizations and others (LinkedIn).

In June 2017, Copenhagen-based shipping giant AP Moller Maersk, who carried one-fifth of the world's freight rate, was shot by "NotPetya Ransomware" Malware as part of a national attack. The virus has affected the company's operations in Rotterdam, Los Angeles, Mumbai, Auckland, and many ports around the world and has caused interruptions and delays in weeks. In this context, all the infrastructures of the company were renewed, that is, 4,000 new servers, 45,000 PCs, 2,500 applications had to be installed. Maersk's attitude in this case was fast and transparent, and it was evaluated by experts that it was trying to help other companies from day one through an open dialogue. This $300 million cyber attack on AP Moller Maersk's digital infrastructure had a triggering effect, and had even been described by the company's president as a very important wake-up call not just for Maersk, but for the entire global supply chain. Special seminars and workshops have increased on the subject (Digital Ship 2:14).

On 24 July 2018, China shipping company China Ocean Shipping (COSCO) was attacked and the event was identified as ransomware. The company's terminal at Long Beach Port was affected by the attack, and their links to other regions were closed and measures were taken to make further investigations (World Maritime News).

4.7 International Evaluations and Sectoral Regulations on Cyber Security

One of the leading international cyber security activities is the implementations of the International Telecommunication Union (ITU), a special agency in the field of information and communication technologies

> *"Shipping is an industry that's hard to change, but new market opportunities and new technologies are making transformation inevitable".*
>
> Shaun Crawford,
> EY Global Vice Chair – Industry

at United Nations. These applications are (i) developing the capacity at national and regional level, (ii) working with member states to assist in the establishment and development of national cyber-incident response teams (CIRTs), (iii) publishing reports stating that countries are graded according to criteria such as international participation in information security, national cyber security strategies and trainings in this field (ITU).

ITU Global Cyber Security Index (GCI), which was calculated by a survey that measured the member states fulfillment of their cyber security commitments, was last published in 2018.

It is currently considered that there are visible gaps in many countries in terms of capacity and programs in the field of cyber security and national cyber security strategies (NCS) and the capacity of computer emergency response teams (CERT) to disseminate strategies and information on the implementation of cyber crime legislation (ITU, 2018: 8).

There are many laws, conventions and guidelines governing maritime transport and setting rules for environment, ship safety, safety and efficiency. These rules and regulations are important for maritime industry and cyber security rules have to be included in maritime contracts. Different approaches should not be taken at local, national, regional and global level in order to take necessary precautions for cyber security and to ensure sustainable cyber resilience. For this purpose, measures against cyber attacks in the maritime industry have been initiated with the publication of various authorities and institutions in the form of recommendations, guidelines and circulars in response to cyber attacks. Some examples to these measures are listed below:

- Strategies are defined and reports are published by regional and national authorities such as USA and EU.
- IMO and various maritime organizations (BIMCO, DNV GL, IACS, LR, CLIA, ICS, INTERCARGO, INTERTANKO, OCIMF and IUMI) develop recommendations, guidelines, class approvals and certificates.

- Oil and gas industry joins forces to tackle cyber threats.
- International Standards of Organization (ISO)/International Electrotechnical Commission (IEC 27001) published Information Security Management System (ISMS) Standards.

European Union's Cyber Security Strategy, which will guide the European Union for short- and long-term strategies, was signed on 2 July 2013. Strategies include cyber flexibility, reduction of cybercrime, development of cyber policies and capabilities; development of industrial and technological resources for cyber security and the creation of a coherent international cyber policy for the European Union (Hayes, 2016: 56).

In 2016, EU has issued the 2016/1148 directive (EU) for the safety of networks and information systems applicable to ports, and the General Data Protection Regulation (EU) 2016/679, which may be applicable for ships from May 2018 (Rossi, 2018: 29).

The National Institute of Standards and Technology (NIST), affiliated to the US Department of Commerce, has published the cyber security framework and road map as a starting point for creating cyber security awareness in all businesses, from micro- to large-scale companies. In addition, "USCG Cyber Strategy" document has been published in 2015 by prioritizing three specific strategic titles. These are defending cyberspace, enabling operations, and protecting infrastructure (USCG, 2015: 9).

IMO has attempted to raise awareness of how to deal with risks in a marine cyber risk management approach to cyber attacks and security breaches, which will cause serious harm to the safety of ships, ports, coastal facilities and other maritime industry units. As a result of the studies initiated in 2014, the "Guidelines on maritime cyber risk management" document was published by IMO in 2016. With the change published in 2017, cyber security is included in the International Ship and Port Facility Security Code (ISPS) and the International Security Management Code (ISM). Companies are allowed to be incorporated in their own Safety Management System (SMS) until 1 January 2021, otherwise they are at risk of being detained (IMO 2).

ISM elaborate how port and ship operators should carry out risk management processes. It is considered that operators will at least be aware of cyber risks to make cyber security an integral part of these processes.

Since the maritime sector and especially the ships are vulnerable to cyber attacks, some institutions within the sector such as BIMCO, CLIA, ICS, INTERCARGO, INTERTANKO, OCIMF and IUMI, have come together and published the Guidelines on Cyber Security Onboard Ships. In this guide generally, the creation of cyber awareness due to cyber security risks; protection of computer and communication technology infrastructure systems in ships; authorizing persons to access the necessary information; protection of information according to the vulnerability on board; management of communication between the ship and the coast; development of a cyber-event plan based on risk assessment principles are discussed (BIMCO 2018).

In addition, the clause brought up by BIMCO for the cyber security risks and incidents that may affect the ability of one of the parties to fulfill their obligations for contracts. The clause will perform two important functions. The first is to increase awareness of cyber risks between owners, charterers and brokers. The second function will provide a mechanism to ensure that the parties to the contract have procedures and systems to help minimize the risk of an incident occurring and mitigate their impact if the incident occurs (BIMCO).

In April 2017, Oil Companies International Marine Forum (OCIMF) published the third version of Tanker Management and Self-Assessment (TMSA), two part of which are related to cyber security. It has been ensured that this requirement is an industrial necessity for the maritime companies operating the tanker. According to the document, companies will be provided with a guide on how to identify and mitigate cyber security, software management procedures, cyber threats, current guidelines for cyber security, password management procedures, and cyber security plan was requested (Lagouvardou, 2018: 28).

In the maritime sector, it was necessary to establish Class approvals/ notations and certificates by classification societies in order to accelerate the establishment of cyber security standards, develop their own frameworks and cyber security solutions, and establish a cyber-safe marine domain.

Based on the IMO directives and other reference standards, it is foreseen to regulate cyber security class notations to help identify the cyber risks with the annual periodic surveys of the ships during the design phase, construction phase and service period of the ships. Another important issue is whether this will be implemented globally efficiently, or if another 'one page is added to the annual audit' mentality. Almost all classification societies in particular IACS (International Association of Classification Classification Societies) and its member organizations are working to develop their own guides, class approvals and notifications, and cyber security solutions. Concisely, the class organizations want to guide the digital development of the industry with their standards and certifications.

In 2018, the International Association of Classification of Societies (IACS) has identified 12 recommendations (Tab. 4.3) on cyber security to

Tab. 4.3: IACS Recommendations (IACS)

1. Software maintenance	7.	Network Security
2. Manual Backup	8.	Vessel System Design
3. Contingency plan	9.	Programmable System Equipment Inventory
4. Network Architecture	10.	Integration
5. Data Assurance	11.	Remote Update/Access
6. Physical Security	12.	Communication and Interfaces

highlight cyber flexibility requirements for ships' operational lives (IACS).

In 2016, USA Classification Society ABS published the document called "ABS Guide for Cybersecurity Implementation for the marine and offshore industries" and updated in 2018. With the class notations defined by the document, and how a company sees the cyber security risk and provides the content of the processes applied to manage this risk for both systems (OT and IT). Notations indicate the level of cyber preparation of ship owners, operators, processes, procedures and assets (ABS, 2016: 17–19).

ABS offers the optional CS series CS1, CS2, CS3, and CS-Ready.

- CS1 Informed CyberSafety Implementation is informal management of risks, policies and procedures. Informal management of the OT and/or IT cybersecurity threats and technology landscape.
- CS2 Rigorous CyberSafety Implementation is formal systematic risk management via global enterprise policies and procedures.

- CS3 Adaptive CyberSafety Implementation (Highest level of Readiness) is formal systematic risk management via global enterprise policies and procedures with demonstrable continuous improvement processes.
- CS-Ready notation focuses on the cyber security of Operational Technology during the construction and delivery of ships (ABS, 2016: 119). The CS-Ready Notation requires the proper characterization of functions and connections. So, the owner can control access to connections and systems (Marinelog).

Italian classification society RINA has started a digital transformation with RINACube, a cloud-based platform that can collect and integrate data and digital assets from various sources.

Subsequently, RINA has launched the Digital Ship Notation, the first additional class notation available for shipowners willing to demonstrate the added value of their ship or fleet through the efficient use of digital technology (RINA).

'Cyber-enabled systems', developed by the UK CLASS organization LR, can be traditionally controlled by the crew but thanks to the latest developments in IT and Operational Technology (OT) it is considered to be the ship systems that can now be monitored or both monitored and controlled either remotely or autonomously (Lloyd's Register, p. 30). Cyber-Safe, Cyber-Maintain, Cyber-Secure and Cyber-Performance are released as four new class notation and certificates that are being introduced to identify the "Autonomy Level" achieved in the areas of "Cyber enablement" (UNECE).

The class society 'DNV GL' issued the class notation 'Cyber secure' that aims to help owners and operators protect their assets from cyber security threats. Cyber Secure class notation provides a basis for addressing cyber security levels for the main functions of a ship during operation or construction. The Cyber Secure class notations have three different qualifiers: Basic; Advanced; and (+) Plus (DNV GL 2).

- "Basic" is intended for ships in operation.
- "Advanced" is intended for newbuild ships, where cyber security will be included and integrated into the vessel design.
- "+ [Plus]" is intended for extra systems that are not part of the standard scope.

> *"Crime is going digital. Be aware, keep updated, demonstrate cyber resilience and stay secure on board".*
>
> Capt. Erol Eryaşa

2.8 Overview

Maritime industry which is the most important infrastructure of international trade has various problems in terms of cyber security. These problems are mainly due to the dynamic changes in technology.

Cyber security is an ever-changing field as new threats and countermeasures emerge, and it is still one of the biggest concerns in the maritime sector. Regarding e-navigation, autonomous ships and even unmanned ships (Remote Controlling) the likelihood of diversity, frequency and complexity of cyber attacks are increasing.

If cyber security awareness is not fundamentally established, the level of cyber threat will continue to rise. Although the issue of cyber security seems to be a package of measures that can be provided by technological infrastructure, it is also a matter of culture and attitude. Therefore, it is necessary to give importance to training to increase the general awareness of cyber risks, to ensure that appropriate behaviors and technical competences are included in maritime management and to help everyone be prepared to develop new skills.

Effective cyber risk management should start at the senior management level. Senior management should place a culture of cyber risk awareness on all levels of an organization and provide a holistic and flexible cyber risk management regime that is continually evaluated through continuous and effective feedback mechanisms (IMO, 2017).

The recommendations, directives and guidelines of leading organizations in the maritime sector, including the classification bodies, especially IMO principles and policies, may have important and positive effects in the fight against future cyber threats as well as the known cyber threats. In addition to the measures taken, it may be useful to prepare the Cyber Code regulation, which will enable the standardization of the measures to be taken against the digitalization of the maritime sector, in accordance with the digitalization levels of the ships and coastal structures.

On the one hand, the most preferred option in global supply chain management is maritime transport and on the other hand it is cyber world where the two meet. There will be a lot of research and development about these topics. At this intersection, many disciplines can take place and experts, academics from different fields will be able to illuminate the depths of maritime and cyber security.

References:

ABS (American Bureau of Shipping). Guide for Cybersecurity Implementation for the Marine and Offshore Industries (September 2016—Updated 15 June 2018).

Balduzzi, M., Pasta, A. and Wilhoit, K. (2014). "A Security Evaluation of AIS Automated Identification System". Proceedings of the 30th Annual Computer Security Applications Conference: pp. 436–445.

BIMCO. (The Baltic and International Maritime Council), (https://www.bimco.org/news/priority-news/20181121-cyber-security-clause) Access Date 02.03.2019.

BIMCO. (2018). "The Guidelines on Cyber Security Onboard Ships", Version 3.

Connecting EU. "How Important Is the Cyber-Security in Maritime Field Today?", https://connecting-eu.onthemosway.eu/2017/05/important-cyber-security-maritime-field-today Access Date 01.04.2019.

Cybersecurity Ventures. (2018). "Cybersecurity Market Report" https://cybersecurityventures.com/cybersecurity-market-report/ Access Date 01.05.2019.

Daum O. (2019). "Cyber Security in the Maritime Sector". *Journal of Maritime Law & Commerce*, 50(1): 1–19.

Digital Ship 1, (www.digitalship.com) "Digital risks in the maritime sector", December 2018–January 2019.

Digital Ship 2 (www.digitalship.com) "Port of Rotterdam puts IoT platform into operation" April–May 2018.

DNV GL (Det Norske Veritas and Germanischer Lloyd) 1: https://www
.dnvgl.com/services/cyber-secure-class-notation-124600 Access Date
19.03.2019.

DNV GL (Det Norske Veritas and Germanischer Lloyd) 2: Einarsson S.
(2018). https://www.dnvgl.com/expert-story/maritime-impact/digital-
defence.html Access Date 05.02.2019.

Fahey, S. (2017). "Combating "Cyber Fatigue" in the Maritime
Domain". *Humanitarian Law & Policy* blog, https://blogs.icrc.org/
law-and-policy/2017/12/07/combating_cyber-fatigue-in-the-maritime-
domain/ Access Date 15.03.2019.

Garcia-Perez, A., Thurlbeck, M. and How, E. (2017). "Towards
Cyber Security Readiness in the Maritime Industry: A
Knowledge-based Approach". *International Naval Engineering
Conference*. https://pureportal.coventry.ac.uk/en/activities/
international-naval-engineering-conference

Gliha D. (2017). "Maritime Cyber Crime—21st Century Piracy". *Anali
Pravnog fakulteta Univerziteta u Zenici*, (20): 228–238.

GMF (Global Maritime Forum). (2018). *Global Maritime Issues
Monitor*. Copenhagen.

Hayes, C. R. (2016). "Maritime Cybersecurity: The Future of
National Security". (Calhoun: The NPS Institutional Archive) Naval
Postgraduate School. Monterey, California, Master Thesis.

Houston Chronicle, https://www.houstonchronicle.com/business/energy/
article/Malware-on-oil-rig-computers-raises-security-fears-4301773.
php Access Date 12.03.2019.

IACS (International Association of Classification of Societies) http://
www.iacs.org.uk/news/12-iacs-recommendations-on-cyber-safety-
mark-step-change-in-delivery-of-cyber-resilient-ships Access Date
02.04.2019.

ICS (International Chamber of Shipping), Shipping and World Trade,
http://www.ics-shipping.org/shipping-facts/shipping-and-world-trade,
Access date 04.03.2019.

IET (Institution of Engineering and Technology). (2017). Code of
Practice. Cyber security for Ships. London-United Kingdom.

IHS Markit. (2018). Maritime Cyber Survey- the results. London.

IMO 1 (International Maritime Organisation): IMO Multilingual
Glossary on Cyberterms. Retrieved from: http://www.gard.no/
Content/21112214/CYBERTERM_Imo.pdf Access Date 13.02.2019.

IMO 2, http://www.imo.org/en/OurWork/Security/Guide_to_Maritime_
Security/Pages/Cyber-security.aspx Access Date 03.03.2019.

IMO. (2015). Resolution A.1106(29) Adopted on 2 December 2015
(Agenda item 10) Revised Guidelines for the Onboard Operational Use
of Shipborne Automatic Identification Systems (AIS). Retrieved from
http://www.imo.org/en/KnowledgeCentre/IndexofIMOResolutions/
Assembly/Documents/A.1106(29).pdf

IMO. (2017). MSC-FAL.1/Circ.3 5 July 2017: Guidelines on
Maritime Cyber Risk Management. International Armour Co.
"Maritime Cyber Security" http://www.armour.gr/catalogues/pdf/
CyberSecurityOnBoard.pdf Access Date 10.03.2019.

IRM. (The Institute of Risk Management). (2014). Cyber Risk Resources
for Practitioners. London.

ITU. https://www.itu.int/en/ITU-D/Cybersecurity/Pages/national-CIRT.
aspx Access Date 12.01.2019.

ITU. (International Telecommunication Union). (2008). Overview of
cybersecurity Recommendation ITU-T X.1205. Switzerland –Geneva.

ITU (International Telecommunication Union.). (2018). (Global
Cybersecurity Index (GCI).

Keleştemur A., Yapıcı, M., Koldemir B. (2017), "Deniz Taşımacılığında
Siber Güvenliği Tehdit Eden Unsurlar ve Koruma Önlemleri
Üzerine Bir Çalışma", (Researchgate) https://www.researchgate.net/
publication/320698188 Access Date: 12.04.2019.

Lagouvardou, S. (2018). "Maritime Cyber Security: Concepts, Problems
and Models". Technical University of Denmark, Department of
management Engineering, Master Thesis.

Linkedin: Chronis Kapalidis (2018). https://www.linkedin.com/pulse/
4-cases-cyber-security-failures-shipping-history-chronis-kapalidis/
Access Date 26.11.2018.

Lloyd's Register. "LR approach to cyber security, Marine and Ofshore".
United Kingdom https://www.rina.org.uk/res/LR%20Approach%20
to%20Cyber%20Security.pdf Access Date 04.04.2019.

Lloyd's Register. (2017). "Cyber-enabled ships, ShipRight procedure assignment for cyber descriptive notes for autonomous & remote access ships", A Lloyd's Register guidance document (Version 2.0, December 2017).

Maersk. https://www.maersk.com/news/2018/06/29/maersk-and-ibm-introduce-tradelens-blockchain-shipping-solution Access Date 02.12.2018.

Marinelog. https://www.marinelog.com/news/abs-issues-first-cyber-security-ready-notation-to-hhi-vlcc/ Access Date 02.03.2019.

Marinemec. https://www.marinemec.com/news/view,bridge-systems-are-not-cybersecure_53333.htm Access Date 18.03.2019.

NewsScientist. https://www.newscientist.com/article/2143499-ships-fooled-in-gps-spoofing-attack-suggest-russian-cyberweapon/ Access Date 03.02.2019.

Pajunen, N. (2017). "Overview of Maritime Cybersecurity". South-Eastern Finland University of Applied Sciences, Bachelor's Thesis Marine Technology, Finland.

ReedSmith. (2019). Digital Age Survey, Is the shipping industry embracing the digital age?

RINA. https://www.rina.org/en/media/press/2018/08/30/gnv-certifies-its-fleet Access Date 15.03.2019.

Rolls-Royce. https://www.rolls-royce.com/media/press-releases/2017/03-10-2017-rr-joins-forces-with-google-cloud-to-help-make-autonomous-ships-a-reality.aspx, Access Date 15.12.2018.

Rossi. (2018). Maritime Cyber Security; https://www.sjofart.ax/sites/www.sjofart.ax/files/attachments/page/maritime_cyber_security_interferry_committee_2018_distrib.pdf.

Safety4sea (1). https://safety4sea.com/cm-blockchain-ventures-in-shipping-where-we-stand/ Access Date 03.04.2019.

Safety4sea (2). https://safety4sea.com/cm-maritime-ready-to-ride-on-internet-of-things/ Access Date 14.04.2019.

Safety4sea (3). https://safety4sea.com/pirates-exploiting-cybersecurity-weaknesses-in-maritime-industry, Access Date 02.02.2019.

Saygılı, M. S. and Ünal A. N. (2018). "Cyber Terrorism Risk at Ports and Organizational Management Process in Application of Security Plan". In Özseven, T. (Ed) and Karaca, V. (Ed), *International*

Symposium on Innovative Approach in Scientific Studies (ISAS Winter 2018) Proceedings, (246–251), Samsun/Turkey.

Sea Knight Maritime Inc. https://seaknightmaritime.com/maritime-security-news/f/how-to-steal-a-ship, Access Date 03.02.2019.

Seatrade http://www.seatrade-maritime.com/news/europe/antwerp-incident-highlights-maritime-it-security-risk.html Access Date 03.03.2019.

Secure State Cyber https://securestatecyber.com/cyberbloggen-en/the-future-of-maritime-cybersecurity/, Access Date 02.02.2019.

Sofronis F. (2018). Maritime Cyber Security. EY (Ernst & Young) London.

Trelleborg Marine Systems. (2018). Use of big data in the maritime industry. Access Date 03.01.2019 https://www.patersonsimons.com/wp-content/uploads/2018/06/TMS_SmartPort_InsightBee_Report-to-GUIDE_01.02.18.pdf.

TrendMicro. https://blog.trendmicro.com/trendlabs-security-intelligence/vulnerabilities-discovered-in-global-vessel-tracking-systems Access Date: 03.02.2019.

UNCTAD. (2017). (United Nations Conference on Trade and Development). *Review of Maritime Transport.*

UNCTAD. (2018). (United Nations Conference on Trade and Development). *Review of Maritime Transport.*

UNECE (The United Nations Economic Commission for Europe). https://www.unece.org/fileadmin/DAM/trans/doc/2018/sc3wp3/07._LR.pdf Access Date 03.04.2019.

USCG (United State Coast Guard). (2015, June). Cyber Strategy. U.S Coast Guard Headquarters Washington, D.C.

World Economic Forum. (2019). The Global Risks Report.

World Finance: Top 5 of the fastest-growing industries in the World https://www.worldfinance.com/markets/top-5-of-the-fastest-growing-industries-in-the-world Access date 08.03.2019.

World Maritime News. https://worldmaritimenews.com/archives/257665/cosco-shipping-lines-falls-victim-to-cyber-attack/ Access Date 08.01.2019.

Zăgan, R., Raicu, G., Hanzu-Pazara, R. and Enache, S. (2017). "Realities in Maritime Domain Regarding Cyber Security Concept". *Advanced Engineering Forum ISSN: 2234–991X*, 27, pp 221–228.

Ahmet Naci Ünal

5. Cyberspace and Chaos: A Conceptual Approach to Cyber Terrorism

Abstract: Internet technology, which entered our lives in the 1990s, was called cyber environment or cyber world. Nowadays, the sensors used in various fields with embedded systems can communicate with other sensors besides Internet connection. This expanding network structure is defined as cyberspace. This process has led to the introduction of cyber prefix in front of many concepts such as cyber resilience, cyber bullying, cyber security and cybercrime. One of these areas is cyber terrorism. In this study, the concepts of terrorism and terrorism are explained first. Then, the concept of cyber terrorism, measures to be taken against cyber terrorism, cyber terrorism threat analysis, cyber threat classification, evaluation of cyber threat effects, cyber deterrence, determination of cyber attacker profile and current cyber threats are examined. Following the general evaluation, chapter ends.

Keywords: Decision Support Systems, Information Systems, Information Management, Cyber Terrorism, Cyber Security

5.1 Introduction

Human beings have felt the necessity to store their crops, starting from the period that we call the *Agricultural Society*, wanting to reach these crops intact after a certain period of time, and started marketing activities by producing various products from these crops. After this process, which lasted for thousands of years, the product was replaced by data, which was defined as the "structural code of a phenomenon" by Vercellis (2009: 6). Especially in the 2000s, as Sanders (2016: 224) showed in Fig. 5.1, data entered into a transformation process.

In this process, data have been processed for a defined purpose or purposes and transformed into information, information has been developed and transformed into knowledge, and using this knowledge, wisdom has been achieved. Within the scope of this transformation, accessing, processing and storing information come to the fore, and the present information is used personally or stored for future use.

In today's world, the rapid increase of the amount of data and the direct effect of this increase on the life of the community leads to increases in

Fig. 5.1: The Journey of Data (Sanders, 2016: 224)

issues or problems faced by individuals. This acceleration requires that the process of solving the problem areas should also be fast. This intensity and the desire to produce quick solutions directly affect the daily lives of societies, particularly those with high and rapid data flow. The more structured this process, the more regular the daily lives of these societies. The changes that will disrupt this order increase the level of anxiety at various degrees based on social strata. This increase in the level of anxiety increases the level of risk, which is a probability-based word used to describe the dangerous situations and concerns that people may face, along with an uncertain impact on individuals. It is almost impossible to liken the risks encountered in the past with the risks emerging today. In the most general sense, the concept of risk is the multiplication of the probability of occurrence of any event with the effect it will generate if realized. Therefore, since all activities are structured in the cyberspace dimension in today's IT world, the impact value can be high even if there is a low probability of occurrence. Therefore, this risk expectation brings along uncertainty. Uncertainty, on the other hand, represents an environment that cannot be fully perceived due to the lack of sufficient data in the hands of the individual and society at large. This new environment inevitably reveals the concept of *chaos* that can be expressed as *irregular and unpredictable behavior of complex, nonlinear dynamic systems*. This process leads to the deterioration of the structured lives of societies and increases their anxiety levels.

One of the factors that will increase the level of social anxiety by increasing the level of risk in our daily lives in today's world is the concept of terrorism. Terror is defined in the *Cambridge Dictionary* as "extreme fear, or violent action that causes fear" while it is defined as "the use of extreme fear to intimidate people" in the *Oxford Dictionary*.

Although the definition of terror is generalized enough to enter into dictionaries, the same is not valid for the concept of terrorism (Tab. 5.1).

Tab. 5.1: Definition of Terrorism by Some Countries (OECD)

	Intention of Terrorist Act	Means Used	Targets/Effects
Australia	Action done or threat made, with the intention of advancing a political, religious or ideological cause, with the intention of coercing or influencing by intimidation of the government of Australia or the Australian States or Territories, or a foreign country, or intimidating the public.	An act (or threat of an act), that is not advocacy, protest, dissent or industrial action, that causes specified damage.	An action that causes serious harm to a person, serious damage to property, causes death or endangers life or creates a serious health or safety risk, or seriously interferes with, or disrupts or destroys an electronic system.
Austria	To influence the government or put the public or any section of the public in fear.	Act or threat of violence.	Human life, tangible or intangible property or infrastructure.
Germany	Acts committed for political, religious, ethnic or ideological purposes suitable to create fear in the population or any section of the population and thus to influence a government or public body.		The insurer shall indemnify, if this has been specially agreed, in respect of insured property which is destroyed, damaged or lost due to: a) fire, explosion, b) impact or crash of aircraft or aerial bodies and vehicles, also craft, of all kinds, their parts or their cargo, c) Other malicious damage, insofar as the mentioned perils are caused by an act of terrorism committed in the Federal Republic of Germany.

Continued on next page

Tab. 5.1: continued

	Intention of Terrorist Act	Means Used	Targets/Effects
United Kingdom	Acts committed for political, religious, ethnic or ideological purposes suitable to create fear in the population or any section of the population and thus to influence a government or public body.	Act of violence	Commercial property and consequent business interruption costs arising from an act of terrorism.
United States	Part of an effort to coerce the civilian population of the United States, or to influence policy or affect the conduct of the US by coercion.	Violent act or dangerous act	Endanger human life, property or infrastructure that results in damages within the United States, or outside the US in the case of an attack of an air carrier or vessel, or premises of a US mission.

Although there are various approaches to the definition of terrorism, it is an undisputed reality that the target of terrorism is human lifestyle and human life. The problems experienced in the definition of terrorism can also be experienced in the classification of terrorism and therefore, it is structured in different ways. In addition, some scientists working on terrorism limit the subject to only their own fields of science, and are moving away from the fact that terrorism is an interdisciplinary concept.

Within this context, the classification compiled from studies conducted by Anatasescu and Voicu (2015: 12–15), Chalk (1996: 12–22), Jenkins (2006: 117–130), Petrevska et al. (2016: 76–77), and Şimşek (2016: 323–326) and cited in many studies is depicted in Fig. 5.2.

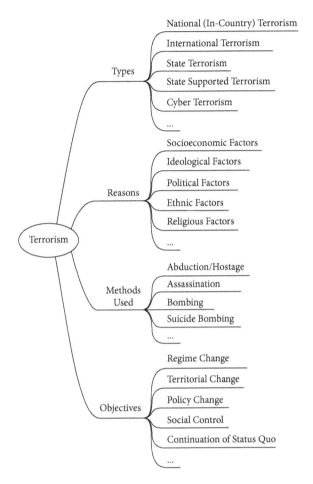

Fig. 5.2: Stages of Terrorism

Furthermore, Chalk (1996: 12–22) states that the main characteristics of terrorism include the fact that it is a political activity and a psychological form of war, such activities affect the civilian population in general, it is systematic in itself, and it aims awareness.

Today, cyber terrorism is almost as significant as terrorism. Although cyber terrorism is shown as a branch of terrorism in Fig. 5.2, they are considered equivalent threat elements in the context of the effect they can create in society.

5.2 What Is Cyber Terrorism?

Denning (2000: 1) defines cyber terrorism as illegal threats and damaging attacks on computers, network systems, information and databases of political, social strata and individuals in order to intimidate and create pressure on such strata and persons. The National Infrastructure Protection Center (NIPC) defines cyber terrorism as "a criminal act conducted with computers and resulting in violence, destruction, or death of targets in an effort to produce terror with the objective of coercing a government to amend its policies." And the FBI defines cyber terrorism as any "premeditated, politically motivated attack against information, computer systems, computer programs, and data which results in violence against non-combatant targets by sub-national groups or clandestine agents" (Alford, 2017: 13). Based on these definitions, it can be said that malicious people aim to harm the masses with cyber terrorism, to weaken the economy, to deteriorate the morale-motivation of people, and to harm national security.

According to Jarvis, Nouri and Whiting (2014: 28), the main goal of cyber terrorism, or in other words a cyber-attack is to differentiate cyber terrorism from other types of attacks motivated for a political purpose. Enney (2015: 1) points out that confusion about cyber terrorism is, in part, caused by attempts by hacktivism and terrorists to broaden the Internet to facilitate traditional terrorist acts.

The sections that make up cyber terrorism and the sub-sections of these sections can change over the years and are expressed in different ways. In this regard, utilizing the studies conducted by Salleh et al. (2016: 1035), Mazari et al. (2016: 11–18), Yalman (2018: 259–279), Ahmad et al. (2012:232), it is possible to reach the six main sections, namely, the objectives, motivation, means, effect, intention and actual methods as shown in Fig. 5.3. However, what should be considered in this segmentation is that cyber terrorism is a dynamic process. Cyber terrorism is similar to a living organism as in other technological systems. For this reason, it constantly renews itself and transforms. Although there are various differences in the scope of the definitions, the tools of attack are the information and communication technologies that have been increasing rapidly in people's daily life (Saygili and Unal, 2018: 247). Therefore, the adapted structure shown in Fig. 5.3 is not a final representation, but only a case finding.

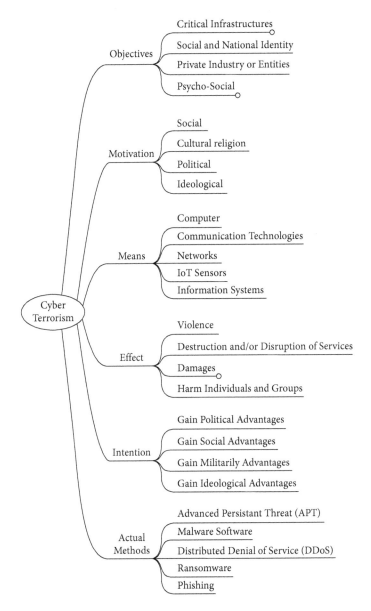

Fig. 5.3: Cyber Terrorism

Here, the areas targeted by cyber terrorism are shown under objectives. The resources used to implement cyber terrorism are in the motivation section. The tools used by cyber terrorism are classified under means. The effect section includes the targeted impacts on society or in the information system through realization of cyber terrorism. The intention section includes the intended objectives using cyber terrorism. The actual methods section includes cyber-attack methods, which are used with the help of technological developments. Another point to be noted here is the ability of conventional terrorist groups to use cyberspace and cyber terrorism itself are different phenomena. When the study of Evan, et al. (2017: 13–17) is examined in this regard, the structuring shown in Fig. 5.4 is encountered.

As depicted in Fig. 5.4, terrorist groups use cyberspace for enabling, disruptive, and destructive purposes. When these three main sections are examined:

- Within the scope of an enabling activity, they can communicate through web pages established by terrorist organizations or terrorist groups, and

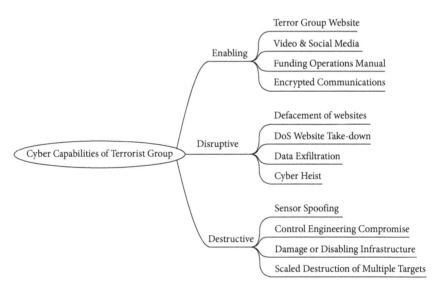

Fig. 5.4: Cyber Capabilities of Terrorist Groups (Evant, 2017: 13–17)

they can conduct illegal propaganda activities with videos and other visual documents. They can also use web-based systems as an encrypted communication tool among themselves.

• Disruptive activity includes activities such as terrorist groups' seizing the targeted web sites, leaving them completely dysfunctional, making them into zombie computers, making them part of DDoS attacks, performing various infiltration activities and cyber theft. Destructive Activity is considered to be the highest level accessible by terrorist elements. The activities of terrorist groups reaching this level can be summarized as sending fake data to secure systems using sensors that connect with cyberspace, infringing and attacking various software-controlled systems and organizing attacks on critical infrastructure facilities and simultaneously organizing attacks on different targets.

5.3 A Conceptual Approach to Counter Cyber Terrorism

It is not possible to talk about a 100 % safe environment in cyberspace and furthermore, all systems are under the influence of the cyberspace. The reason for this is that the threat elements constantly develop and increase the risk level. Cyber threats are the most significant factors that increase this risk level. A detailed risk analysis is needed to reduce the risk level. In order to make this analysis healthy, it is first necessary to determine the existence of the threat. Therefore, it is necessary to recognize the threat, to identify the threat by analyzing the recognized threat data and to determine the type of threat and the attack phase. In other words, the following questions about the threat should be answered:

– What is it?
– Where is it?
– What does it do?

When the threat comes from the cyberspace dimension, it becomes difficult to find answers to the questions mentioned, and the time for the threat determination is prolonged. The key concepts in recognizing cyber threats include cyber threat analysis, cyber-threat classification, cyber-threat impact assessment, cyber deterrence, and identification of cyber-aggressive profiles.

5.3.1 Threat Analysis in the Scope of Cyber Terrorism

We have already mentioned that cyber threats are constantly developing and evolving. Therefore, one of the most important impacts in identifying cyber threats is proportional to developing cyber-threat-based thinking. For this reason, the "Cyber Attack Life Cycle" developed by cyber security researchers, used effectively and shown in Fig. 5.5, is an important turning point.

This figure illustrates the steps taken by attackers against entities. The first step after the target of the cyber-attack is determined, or in the determination phase is to gather information about the target system. This process is conducted by using methods such as port scanning, social engineering, traffic monitoring, and so on. In this way, the weakness areas of the system to be attacked are determined. Once the weaknesses of the target system have been identified, the strategy determination for the acquisition of the target system will begin. At this stage, which cyber-attack tools such as viruses, Trojans, worms, zero day attacks, phishing attacks, ransomware to be used are decided upon. Upon this decision, the target system is attacked. Following the cyber attack, it is necessary not to leave traces in the target system, because leaving a trace means expediting identification of the anomaly in the target system. For this reason, cyber tracks are attempted to be destroyed as effectively as possible. In the final stage, whether the benefit obtained as a result of this attack is productive is investigated. Once

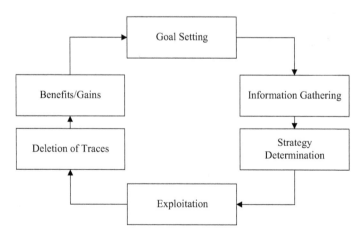

Fig. 5.5: Cyber Attack Life Cycle (Erol and Sagiroglu, 2018: 115)

the cycle is completed as such, the search for a new target will begin. The proper evaluation of this cycle will be useful in obtaining important clues during spotting, identifying, and tracing the cyber attacker.

5.3.2 Classification of Cyber Threats

In order to identify any issue or problem, we need to go down to the root causes of that issue or problem and from that point on, we should move upwards with scientific analysis methods. In other words, we need to properly model or classify the process between the root cause and problem. Thanks to this classification, we can concentrate our attention on the event, as we will narrow the area we will focus on, and we can plan the energy we will use properly. The process that can be the basis for this planning was modeled by Bucci (2009: 2) and classified as shown in Fig. 5.6. This classification is called the Cyber Threat Spectrum.

While this spectrum is being created, scaling involves classifying the danger levels of cyber threats as low and high. In this context, individual hackers are considered to be the lowest-level cyber threat in this seven-stage cyber threat classification, while the nation-state cyber-enabled kinetic

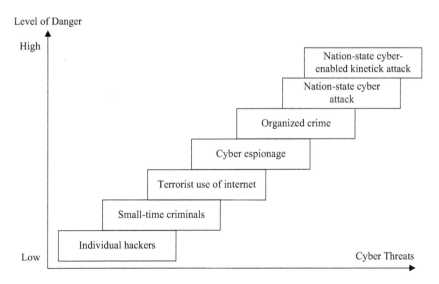

Fig. 5.6: Cyber Threat Spectrum (Bucci, 2009: 2)

attack classification is considered to be of the highest level. The most important factor that should not be ignored in this consideration is that each of these classified threats can select individuals, institutions or states as targets. Of course, the ability to measure cyber-attack effects or cyber-attack activity is as important as the classification or scaling of cyber threats.

5.3.3 Evaluation of Cyber Threat Effects

The human brain evaluates the phenomena around it in three dimensions and when analyzing any problem, it also shapes the solution set in a three-dimensional environment. For this reason, in the definition of cyber-attack activity made by Libicki (2009: 60), a three-dimensional cyber-attack space was used as shown in Fig. 5.7.

The three-dimensional cyber-attack space is seen to consist of the axes of "intensity of attack", "evolution of time", and "duration of attacks". Therefore, the effectiveness of the attacks in this space can be evaluated differently in three dimensions as shown in Fig. 5.7.

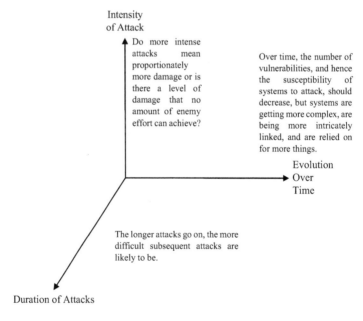

Fig. 5.7: Three Dimensions of the Efficacy of Cyberattacks (Libicki, 2009: 60)

For example, let us assume that the letters A, B, and C shown in Fig. 5.8 represent three different cyber attacks. Here, when we look at the traces of the cyber threats A, B and C in all three dimensions, we see that the effect values are different. For example, it is seen that in the Intensity of Attack dimension B1 > A1 > C1, in the Evolution Over Time dimension C2 > B2 > A2, and in the Duration of Attacks dimension C3 > B3 > A3. In other words, the cyber-attack source B has the most significant effect on the intensity of attack dimension, while the cyber-attack source C has the greatest impact on Evolution Over Time and Duration of Attacks dimensions. Therefore, it is of great importance to classify the differences in the threat assessment phase. The speed of our assessment of this threat and the speed of our response time will be directly proportional to the level of deterrence we have.

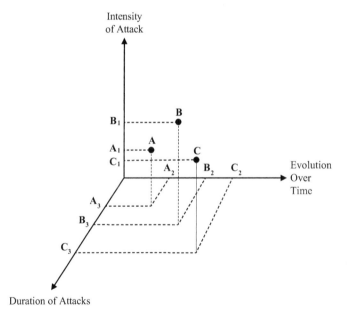

Fig. 5.8: Sample Impact Assessment

5.3.4 Evaluation of the Level of Cyber Deterrence

The concept of deterrence is the ability to neutralize the opposing side without making any attacks with the means we have. Although this concept was mostly used in the military literature until the 2000s, cyberspace's domination of the physical environment, especially after the 2000s, revealed the concept of cyber deterrence.

The study conducted by Libicki (2009: 29) on deterrence, also including cyber capabilities, is depicted in Fig. 5.9.

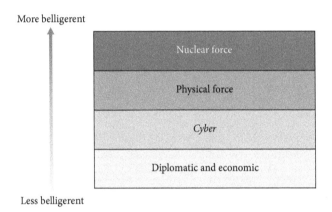

Fig. 5.9: Responses by Rough Order of the Level of Belligerence (Libicki, 2009: 29)

In this classification, diplomatic and economic initiatives take place in the lowest level, while cyber skills exist on it. On this layer, it is seen that there is physical force, and at the top, there is nuclear power. Therefore, according to this scale, the warrior capabilities of the countries with nuclear power are at the highest level. Six years after this study that was conducted in 2009, Bendiek and Metzger (2015: 11) conducted another study, which is shown in Fig. 5.10.

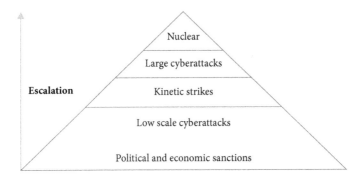

Fig. 5.10: A Possible Model of Escalation (Bendiek and Metzger 2015: 11)

When the graph shown in Fig. 5.10 is examined, it is seen that nuclear power is still naturally located at the top, while cyber attacks are located above and below the kinetic pulse. This shows that a large-scale cyber attack has become almost equivalent to nuclear power in today's world. In fact, because of the long-term effect they will create if they are used, nuclear weapons continue to act only as a deterrent while a large-scale cyber attack can be effective in many areas. It is also easier to control nuclear power since it is under the control of the states. However, cyber attacks can be easily performed by cyber attackers at different levels as shown in Fig. 5.6 (the cyber spectrum) as well. Therefore, the identification of cyber-aggressive profiles in combating cyber terrorism is of great importance, too.

5.3.5 Evaluation of Profiles of Cyber Attackers

Identifying the profiles of the attackers who plan and perform cyber attacks is also an important aspect. One of the important studies on this subject is the study by Rogers (2006: 98) where he proposed nine different profiles based on four different types of attackers on a two-dimensional circumplex circle model. The types here are financial, notoriety, curiosity, and revenge. These eight fields are determined as follows:

- Novice (NV),
- Cyber-punks (CP),
- Internals (IN),

- Petty Thieves (PT),
- Virus Writers (VW),
- Old Guard hackers (OG),
- Professional Criminals (PC),
- Information Warriors (IW),
- Political Activist (PA).

Fig. 5.11 shows the above-mentioned four types and eight profiles on the circle circumplex.

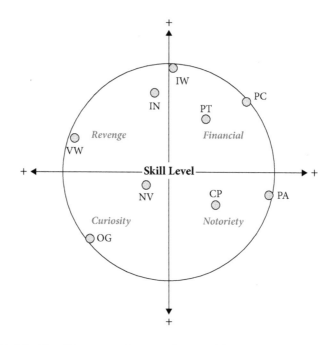

Fig. 5.11: The Two-Dimensional Circumflex Model (Rogers, 2006: 100)

In Fig. 5.11, the technical ability of the attacker increases as the distance away from the center increases, and each circle of the circumplex shows a category of motivation. The technical capability of the attacker increases as s/he moves away from the center within the circle.

5.3.6 Evaluation of Current Cyber Threat Techniques

The most general classification of cyber threats can be made in the form of viruses, worms and Trojans. As the years go by, the contents of these basic harmful objects, their ways of harm and degree of impact increased and they are used with different techniques. These techniques have evolved over the years, resulting in the transformation of cyber threats. The transformation processes of these threats over the years are as shown in the graph in Fig. 5.12.

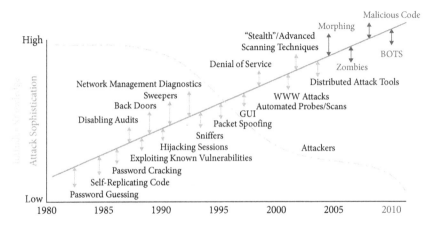

Fig. 5.12: The Increasing Complexity of Threats as Attackers Proliferate (IAEA, 2011: 38)

When this graph is examined, it is seen that attackers had a high level of knowledge in the 1980s but attack sophistication values were low. However, as the years progressed, these evaluations changed in an inversely proportional manner. In other words, hackers have become able to perform highly effective cyber attacks with lower level of knowledge. For example, the data generated by ENISA (2019: 9) based on 2017 and 2018 malware rankings are shown in Tab. 5.2.

Tab. 5.2 shows that even the rankings of the threats used each year may change among themselves. When the threats here are examined, we see that the first four types of attacks remain valid, threats number five

Tab. 5.2: Top Cyber Threats (ENISA, 2019: 9)

Top Threats 2017	Top Threats 2018	Change in ranking
1. Malware	1. Malware	Same
2. Web-Based Attacks	2. Web-Based Attacks	Same
3. Web Application Attacks	3. Web Application Attacks	Same
4. Phishing	4. Phishing	Same
5. Spam	5. Denial of Service	Going up
6. Denial of Service	6. Spam	Going down
7. Ransomware	7. Botnets	Going up
8. Botnets	8. Data Breaches	Going up
9. Insider Threat	9. Insider Threat	Same
10. Physical Manipulation/ Damage/Theft/Loss	10. Physical Manipulation/ Damage/Theft/Loss	Same
11. Data Breaches	11. Information Leakage	Going up
12. Identity Theft	12. Identity Theft	Same
13. Information Leakage	13. Cryptojacking	NEW
14. Exploit Kits	14. Ransomware	Going down
15. Cyber Espionage	15. Cyber Espionage	Same

and six mutually changed places, the Ransomware threat fell to the fourth rank and the cryptojacking technique entered in the list in 2018. The Cryptojacking concept listed here is just like botnet information systems, which can be defined as the unauthorized use of someone else's computer to produce crypto money. Here, just like creating a botnet, malicious software is sent to the information system to be attacked and unauthorized data mining is performed on that computer.

5.4 General Evaluation

The question of how we will be safe in such an intense threat environment is on the agenda. In this context, it is useful to re-examine Fig. 5.3, which shows the lower sections of cyber terrorism. The target of cyber terrorism people and cyberspace based systems that make people's lives easier.

In today's world, the age of reaching cyberspace covers a wide range, starting from the pre-school period and covering the time until people die. For this reason, individuals, societies, organizations, states, in short, all kinds of actors need to determine their cyberspace utilization policies, threat assessments, and behavioral patterns.

It should also be noted that the measures to be developed in cyberspace require an interdisciplinary approach that includes almost all branches of science. Although cyber techniques and technologies are developed by engineers, it should be remembered that end users of these products are people of all demographic groups. In addition, any type of activity in cyberspace triggers a process and these processes are integrated into other processes and continue on their own way.

Therefore, the Cyber Security Life Cycle Framework (NIST-1, 2018: 7–8) shown in Fig. 5.13 consists of five sections: identify, protect, detect, respond and recover.
The contents of these sections are summarized below:

• Identify

Includes development of people, systems, assets, data, and capabilities with an institutional understanding to manage cyber security risk.

• Protect

Includes limiting the impact of a potential cyber security incidents to ensure the delivery of critical services or developing and implementing the

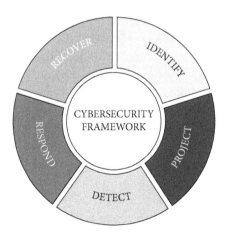

Fig. 5.13: Cyber Security Life Cycle (NIST-2, 2018)

necessary measures. In order to achieve this, access to digital and physical assets should be controlled, processes should be established to secure the data and protective technologies should be used.

• Detect

This dictates a rapid identification of cyber security violations. The detection process involves the timely recognition of anomalies that constitute cyber security events.

• Respond

Refers to the development and implementation of appropriate measures to take action against a detected cyber security event. To this end, a response plan should be prepared, friendly communication lines should be identified, information about the activities should be collected and analyzed, and all necessary activities should be carried out to eliminate the malicious event.

• Recover

Includes developing and implementing appropriate activities to restore all capabilities or services that have been corrupted due to a cyber security incident.

It should be kept in mind that cyber terrorists realize their activities by using the opportunities of cyberspace. Therefore, the measures that should be taken in addition to the cyber security life cycle summarized below are composed of individual and corporate cyber security practices. Within this framework, it is of great importance that institutions keep their cyber security strategies and cyber security implementation policies up-to-date, and create short-, medium- and long-term cyber security plans within the scope of these policies and strategies.

At the very beginning of these plans, different levels of cyber security trainings should be designed and implemented in a practical way to ensure cyber security consciousness and awareness in all layers of society, and to provide interaction between those layers. It should be kept in mind that all software required for sensitive technology, including critical infrastructure hardware and software, must contain national source codes and that it should be written in accordance with the current software-building standards.

The fact that cyber-attacks today are mostly carried out with artificial intelligence software can cause these infiltration activities not to be detected in time by target information systems. In this context, it would be appropriate to develop projects in order to increase the "speed of threat detection" by prioritizing the use of artificial intelligence methods in targeting information system protection as well. It should be noted that in all activities in cyberspace, the weakest link is usually the human being and original projects should be realized related to subjects such as cyber-threat analysis, cyber-threat classification, cyber-threat impact assessment, cyber deterrence, and identification of cyber-aggressive profiles. Please do not forget that cyberspace is almost like a living system. It constantly updates and transforms itself and its effects.

References

Ahmad, R., Yunos, Z., Sahib, S. and Yusoff, M. (2012). "Perception on Cyber Terrorism: A Focus Group Discussion Approach". *Journal of Information Security*, 3(3): 231–237. doi: 10.4236/jis.2012.33029. https://www.scirp.org/journal/PaperInformation.aspx?PaperID=21346 [20.05.2019].

Alford, C. L. (2017). "The Department of Defense Effort to Countering the Cyberterrorism Threat: Is the Threat Real or Hyperbole?". National Defense University Joint Forces Staff College Joint Advanced Warfighting School, Master's Thesis, Norfolk, USA.

Anastasescu, A. and Voicu, G. (2015). "Perceptions Related to Terrorism Phenomenon—Generic Taxonomies". *European Journal of Public Order and National Security*, 7(2): 11–16.

Bendiek, A. and Metzger, T. (2015). Deterrence Theory in the Cyber-century. Lessons from a State-of-the-art Literature Review. Working Paper RD EU/Europe, 2015/02, May 2015 SWP Berlin. https://www.swp-berlin.org/fileadmin/contents/products/arbeitspapiere/Bendiek-Metzger_WP-Cyberdeterrence.pdf.

Bucci, S. P. (2009). "The Confluence of Cyber Crime and Terrorism". Heritage Lectures, Pg.2 https://www.heritage.org/defense/report/the-confluence-cyber-crime-and-terrorism [18.09.2019].

Cambridge (2019): https://dictionary.cambridge.org/tr/s%C3%B6zl%C3%BCk/ingilizce/terror [14.11.2019].

Chalk, P. (1996). *West European Terrorism and Counter-Terrorism the Evolving Dynamic*. London: MacMillan Press Ltd.

Denning, D. E. (2000). Cyberterrorism. The Terrorism Research Center. https://faculty.nps.edu/dedennin/publications/Testimony-Cyberterrorism2000.htm [14.11.2019].

ENISA. (2019). ENISA Threat Landscape Report 2018, https://www.enisa.europa.eu/publications/enisa-threat-landscape-report-2018/at_download/fullReport pg 09. [5.04.2019].

Erol, E. and ve Sağıroğlu, Ş. (2018). "Siber Güvenlik Farkındalığı, Farkındalık Ölçüm Yöntem ve Modelleri". In Sağıroğlu, Ş., Alkan, M. (Ed.), *Siber Güvenlik ve Savunma Farkındalık ve Caydırıcılık*, (115). Ankara: Grafiker Yayınları.

Evan, T., Leverett, E., Ruffle, S. J., Coburn, A. W., Bourdeau, J., Gunaratna, R., and Ralph, D. (2017). Cyber Terrorism: Assessment of the Threat to Insurance. Cambridge Risk Framework series; Centre for Risk Studies, University of Cambridge.

IAEA (International Atomic Energy Agency). (2011). Computer Security at Nuclear Facilities Reference Manual. IAEA Nuclear Security Series No. 17, p. 38. Austria https://www-pub.iaea.org/MTCD/Publications/PDF/Pub1527_web.pdf [14.11.2019].

Jarvis, L., Nouri, L. and Whiting, A. (2014). Understanding, Locating and Constructing Cyberterrorism. In Chen, T. M., Jarvis, L. and Macdonald, S. (Eds.), *Cyberterrorism: Understanding, Assessment, and Response*, (28). London: Springer.

Jenkins, B. M. (2006). The New Age of Terrorism. McGraw-Hill Companies Inc. https://www.rand.org/pubs/reprints/RP1215.html [17.05.2019].

Jensen, R. V. (1987). "Classical Chaos". *American Scientist*, 75(2) (March–April 1987): 168–181.

Kenney, M. (2015). "Cyber-Terrorism in a Post-Stuxnet World". *Orbis*, 59(1): 111–128. http://www.sciencedirect.com/science/article/pii/S0030438714000787 [17.05.2019].

LIBICKI, M. C. (2009). "Cyberdeterrence and Cyberwar". Rand Corp. https://www.rand.org/content/dam/rand/pubs/monographs/2009/RAND_MG877.pdf [18.04.2019].

Mazari, A. A., Anjariny, A., Habib, S. and Mazari, E. N. (2016). "Cyber Terrorism Taxonomies: Definition, Targets, Patterns and

Mitigation Strategies". Proceedings of the European Conference on e-Learning; 2015, pp. 11–18, 8p. http://eds.b.ebscohost.com/eds/pdfviewer/pdfviewer?vid=1&sid=4a3d4ea9-4b64-4d3a-b4dc-daf7dcf5c2ee%40pdc-v-sessmgr03 [15.11.2019].

NIST-1. (2018): "Framework for Improving Critical Infrastructure Cybersecurity" https://nvlpubs.nist.gov/nistpubs/CSWP/NIST. CSWP.04162018.pdf [20.05.2019].

NIST-2. (2018): "Framework for Improving Critical Infrastructure Cybersecurity" https://www.nist.gov/cyberframework [20.05.2019].

OECD. Definition of Terrorism by Country in OECD Countries, https://www.oecd.org/daf/fin/insurance/TerrorismDefinition-Table.pdf [10.04.2019].

Oxford. (2019). https://en.oxforddictionaries.com/definition/terror [15.11.2019].

Petrevska, L., Bulatovic, M., Petrevska, I. and Petrevska, L. (2016). "Plurality of Definitions and Forms of Terrorism Through History". *International Journal of Economics & Law*, 6(18): 71–78.

R. Ahmad, Z. Yunos, S. Sahib and M. Yusoff. (2012). "Perception on Cyber Terrorism: A Focus Group Discussion Approach". *Journal of Information Security*, 3(3): 231–237. doi: 10.4236/jis.2012.33029.

Rogers, M., K. (2006). "A Two-dimensional Circumplex Approach to the Development of a Hacker Taxonomy". *Digital Investigation*, 3: 97–102. https://www.sciencedirect.com/science/article/pii/S1742287606000260 [20.05.2019].

Salleh, N., M., Selamet, S. R., Yusof, R. and Sahib, S. (2016). "Discovering Cyber Terrorism Using Trace". *International Journal of Network Security*, 18(6): 1034–1040. http://ijns.femto.com.tw/contents/ijns-v18-n6/ijns-2016-v18-n6-p1034-1040.pdf [17.05.2019].

Sanders, J. (2016). "Defining Terms: Data, Information and Knowledge". SAI Computing Conference 2016, July 13–15, London, UK. https://ieeexplore.ieee.org/stamp/stamp.jsp?arnumber=7555986 [08.04.2019].

Saygili, M. S. and Unal A. N. (2018). "Cyber Terrorism Risk at Ports and Organizational Management Process in Application of Security Plan". Özseven, T. (Ed.) and Karaca, V. (Ed.), *International Symposium on Innovative Approaches in Scientific Studies (ISAS Winter 2018) Proceedings*, (246–251). Samsun, Türkiye: SETSCI.

Simsek, M. (2016). "Terörizm: Kavramsal Bir Çalışma". Akademik Bakış Dergisi, Sayı: 54: Syf.319–335.

NIST. (2018). "Framework for Improving Critical Infrastructure Cybersecurity" https://nvlpubs.nist.gov/nistpubs/CSWP/NIST. CSWP.04162018.pdf [17.05.2019].

Unal, A. N. (2015). Siber Güvenlik ve Elektronik Bileşenleri: Nobel Akademik Yayıncılık.

Vercellis, C. (2009). Business Intelligence: Data Mining and Optimization for Decision Making. West Sussex: John Wiley & Sons Ltd.

Yalman, Y. (2018). "Siber Terör, Terörizm ve Mücadele". In Sağıroğlu, Ş., Alkan, M. (Ed.), *Siber Güvenlik ve Savunma Farkındalık ve Caydırıcılık*, (259–279). Ankara: Grafiker Yayınları.

List of Figures

Hatice Necla Keleş
New Skills and Talents for the Cyber World
Fig. 1.1:   Workforce Skills Model 16
Fig. 1.2:   Workforce Skills Model Percentage of Time Spent on
            Cognitive Skills 17

Mehmet Sıtkı Saygılı
Supply Chain and Logistics Management in Cyberspace
Fig. 2.1:   Stages of the Industrial Revolutions 23
Fig. 2.2:   Cyber-Physical Systems Components 25
Fig. 2.3:   Big Data Sources and Target Use 26
Fig. 2.4:   Radio Frequency Identification System 34
Fig. 2.5:   Risk Calculation 37
Fig. 2.6:   An Example of a Questionnaire to Evaluate Suppliers'
            Safety Practices and Standards 39

Ahu Ergen
Smart Retailing in Cyberspace
Fig. 3.1:   A Framework for Analyzing Innovations in Shopper
            Marketing 51
Fig. 3.2:   Tailing Pyramid 52
Fig. 3.3:   Factors Identifying a Smart Technology for Retailing 54
Fig. 3.4:   Smart Retailing Framework 55
Fig. 3.5:   Future Shopping Scenario Within the Smart Retailing
            Perspective 56
Fig. 3.6:   Nike Id 59
Fig. 3.7:   Bonobos 60
Fig. 3.8:   McDonalds 61
Fig. 3.9:   Dunkin' Donuts 61
Fig. 3.10: Starship Self-delivery Robot 62
Fig. 3.11: Shopper Demographic Recognition Using In-store Video 64
Fig. 3.12: Shopper Gender Recognition Using In-store Video 65

Ahmet Naci Ünal
Cyberspace and Chaos: A Conceptual Approach to Cyber Terrorism
Fig. 5.1:   The Journey of Data 104
Fig. 5.2:   Stages of Terrorism 107
Fig. 5.3:   Cyber Terrorism 109

Fig. 5.4:   Cyber Capabilities of Terrorist Groups 110
Fig. 5.5:   Cyber Attack Life Cycle 112
Fig. 5.6:   Cyber Threat Spectrum 113
Fig. 5.7:   Three Dimensions of the Efficacy of Cyberattacks 114
Fig. 5.8:   Sample Impact Assessment 115
Fig. 5.9:   Responses by Rough Order of the Level of Belligerence 116
Fig. 5.10: A Possible Model of Escalation 117
Fig. 5.11: The Two-Dimensional Circumflex Model 118
Fig. 5.12: The Increasing Complexity of Threats as Attackers
            Proliferate 119
Fig. 5.13: Cyber Security Life Cycle 121

List of Tables

Hatice Necla Keleş

New Skills and Talents for the Cyber World

Tab. 1.1: Core Work-Related Skills 15

Abu Ergen

Smart Retailing in Cyberspace

Tab. 3.1: Key Differences Between e-Retailing and Smart Retailing 53

Erol Eryaşa

Cyberspace Impacts on Maritime Sector

Tab. 4.1: IT & OT Systems and Their Risks 76
Tab. 4.2: Categorizing the Threats 82
Tab. 4.3: IACS Recommendations (IACS) 93

Ahmet Naci Ünal

Cyberspace and Chaos: A Conceptual Approach to Cyber Terrorism

Tab. 5.1: Definition of Terrorism by Some Countries (OECD) 105
Tab. 5.2: Top Cyber Threats 120